Sustainable Practices for Plant Disease Management in Traditional Farming Systems

Sustainable Practices for Plant Disease Management in Traditional Farming Systems

H. David Thurston

Westview Press
BOULDER • SAN FRANCISCO • OXFORD

Oxford & IBH Publishing Co. Pvt. Ltd.
NEW DELHI • BOMBAY • CALCUTTA

To my wife and parents

Photos are by the author unless otherwise attributed.

Published in 1992 in the United States of America by Westview Press, Inc., 5500 Central Avenue, Boulder, Colorado 80301-2847, and in the United Kingdom by Westview Press, 36 Lonsdale Road, Summertown, Oxford OX2 7EW

Published in 1992 in India by Mohan Primlani for Oxford & IBH Publishing Co. Pvt. Ltd., 66 Janpath, New Delhi 110 001

Library of Congress Cataloging-in-Publication Data
Thurston, H. David
Sustainable practices for plant disease management in traditional farming systems / by H. David Thurston
p. cm.
Includes bibliographical references and index.
ISBN 0-8133-8363-3
1. Phytopathogenic microoganisms—Control. 2. Plant diseases. 3. Sustainable agriculture. I. Title. II. Title: Traditional farmer systems.
SB731.T48 1992
632′.9—dc20 91-15953
CIP

ISBN 81-204-0653-2 (India)

Printed and bound in the United States of America

The paper used in this publication meets the requirements of the American National Standard for Permanence of Paper for Printed Library Materials Z39.48-1984.

10 9 8 7 6 5 4 3 2 1

Contents

CULTURAL PRACTICES

Planting

Land Preparation Practices

Tables and Figures

Tables

Figures

24.1 Seventeen different types of beans grown by a farmer on a one and one-half hectare farm near Puebla, Mexico. The diversity of varieties gave protection against pests and the vagaries of climate. (Courtesy American Phytopathological Society)

24.2 A rain forest in South America. Much of the area in tropical forests is being rapidly lost to cattle pastures.

Preface

This book is only a small beginning and can hardly serve as a review of the immense literature on traditional agriculture, even if we narrow our focus to material relevant to the management of plant diseases. I rapidly discovered, as I began working on the book, that the book would never be finished if I attempted a comprehensive treatment of the subject. Also, much literature in languages other than English and Spanish could not be easily consulted, which seriously restricted the scope of the book.

The vast majority of the information on traditional agriculture pertinent to the management of plant diseases has never been recorded in a form easily accessible to today's farmers or scientists. Most treatments of the history of plant pathology found in textbooks begin seriously with the German Anton de Bary (1831-1888) and consider little of pre-history and ancient writings on plant disease. With rare exceptions those who long ago studied indigenous and traditional agriculture seldom considered or even mentioned plant diseases. The references cited in this book cover a wide range of sources. Some deal directly with the management of plant disease by traditional farmers, others describe traditional farming systems without directly addressing plant disease management, and another large group of references presents research on "modern" agricultural practices that provide insights into mechanisms for disease management that may be operative in traditional systems. This latter information is of importance because of the paucity of formal pathological research within traditional systems.

The study of traditional management of plant diseases should be a rewarding area for future plant pathologists who are not completely seduced by the terms "new" and "innovative" and the prestigious and intellectually appealing basic research in biotechnology. If the 10,000 years of farming experience embodied in traditional agriculture and the knowledge of indigenous peoples can be combined with modern agricultural science, perhaps more sustainable, environmentally sound, and useful agricultural development will result. The book is written for those interested in improving agriculture and the environment.

The emphasis of the book is on plant disease management. Although the terms "pest management" and "pest control" are sometimes used interchangeably, the term "management" is used in this book rather than the term control. Control customarily suggests efforts to eliminate or kill pests, whereas management implies manipulating or managing the farming system so that pests are reduced or remain at non-injurious levels (Thomason and Caswell 1987). In other words, managing pests implies "living with the pest" rather than completely eradicating it, and accepting low levels of damage rather than complete control. The term "biological control" will be retained because of its widespread, common usage. The term "pest" includes plant pathogens (fungi, bacteria, viruses, nematodes), insects, weeds, birds, and rodents. Worldwide, the losses from pathogens, insects, and weeds were estimated at from 15 to 35%, while losses due to diseases alone were estimated at 12% (Cramer 1967).

Several books have been of especially great value in writing this book. First, Palti's (1981) *Cultural Practices and Infectious Crop Diseases* is an excellent, well-organized treatment of cultural practices. Secondly, for anyone interested in the history of plant pathology and traditional plant disease management, Orlob's (1973) treatise on *Ancient and Medieval Plant Pathology* is a rich source of information. I have relied heavily on both of these books for guidance and literature sources. Wilken's (1987) book *Good Farmers: Traditional Agricultural Resource Management in Mexico and Central America* gives excellent insights into the knowledge and practices of traditional farmers in Latin America. The agricultural literature of the ancient Arabs has not been adequately studied by modern scientists. Sezgin (1971) gives an extensive list of references to pre-modern Arabian literature.

Another useful book is Weatherwax's (1951) *Indian Corn in Old America*. His chapter entitled "Historical Sources" is a rich source of references to descriptions of agriculture by various chroniclers shortly after the Spanish conquest of much of the Americas. Priests, soldiers, government officials, colonists, and a few native Americans wrote firsthand accounts of their observations. Weatherwax noted that some of these writers had helped to destroy Indian art and literature and had subjected them to the Inquisition, while others fought for decent treatment and rights for the Indians. He added that, while some writers may have been accomplished liars when it fitted their purposes, falsifying information about agriculture would gain them little. Authors such as De Sahagún (1969), writing in the sixteenth century, used native informants extensively and systematically interviewed them in considerable detail about agriculture.

Palti (1981) has emphasized that the purpose of farming is not the prevention of disease but rather the raising of crops. Farmers, for a variety of reasons, often use practices that exacerbate plant diseases.

Such undesirable practices are found in traditional systems and their use makes an understanding of many traditional practices difficult. Most of this book is concerned with cultural practices for disease management. Analysis of traditional systems and the practices used for disease management used therein suggest that cultural practices predominate strongly. It is quite simple to apply a pesticide or utilize a high-yielding resistant variety to manage plant diseases, but one has to know a great deal about the biology of a situation in order to use cultural management. Thus, the simplicity of chemical and genetic controls has a strong appeal to most farmers.

Common names for plants follow Purseglove (1968, 1972), while common names for diseases and scientific names for pathogens, with few exceptions, follow Holliday (1989). Insects are considered primarily as vectors.

There is considerable overlap in several of the chapters and occasional repetition, which allows chapters to "stand alone." For example, there is some duplication of material in the chapters on fallow and rotation as the two practices are often used together.

Since I was not aggressive or persuasive enough to convince funding organizations of the value of writing this book, the writing, typing, and checking of references was done by the author, and I am solely responsible for any errors.

H. David Thurston

Acknowledgments

The support of the Department of Plant Pathology and the International Agriculture Program of Cornell University is acknowledged. A sabbatic leave in the Section of Plant Pathology of the Department of Agronomy of the Escuela Técnica Superior de Ingenieros Agrónomos of the University of Córdoba in Spain made writing of the final draft possible. The help and hospitality during my sabbatic of Professors Rafael Jimenez Diaz, Miguel Angel Blanco Lopez, and Antonio Trapero Casas is gratefully acknowledged. The sponsorship of the sabbatic by the Dirección General de Investigación Cientifica y Tecnica in Madrid, is also gratefully acknowledged.

I wish to express gratitude for help, advice, or for reading portions of the manuscript to: George Abawi, Phil Arneson, Carmen Barcelo, Tom Barker, John Beer, Jeffrey W. Bentley, Gary Bergstrom, Miguel Angel Blanco Lopez, Merideth Bonierbale, Conrad Bonsi, J. Artie Browning, M. B. Callaway, J. J. Castaño, W. R. Coffman, Rachel Davis, Lourdes Diaz Trechuelo, Carl J. Eide, Peter Ewell, Dennis Finney, Diane Florini, Javier Franco, Charles A. Francis, Edward R. French, José J. Galindo L., Roberto Garcia Espinosa, Stephen R. Gliessman, S. K. Hahn, Yaynu Hiskias, Robert Hoopes, George Hudler, Rafael Jimenez Diaz, Roger Kirkby, Ana Labarta, Rosemary Loria, J. Carlos Lozano, Barbara Lynch, Maria Mayer de Scurrah, Michael Milgroom, Neil Miller, Barbara Mullin, Eric Nelson, John S. Niederhauser, Joanne Parker, Robert L. Plaisted, Alison Power, Gordon Prain, R. Rodriguez-Kabana, Cándido Santiago Alvarez, Gail Schumann, Wayne Sinclair, Steven Slack, Margaret Smith, Christine Stockwell, James Teri, David H. Timothy, Antonio Trapero Casas, Peter Trutmann, Arnold Tschantz, Allen Turner, Madison Wright, David Yohalem, Eldon Zehr, Robert Zeigler, Tom Zitter, Larry Zuidema.

Portions of Chapter 14 on Raised Beds were written together with Dr. Joanne Parker.

The support and patience of my wife, Betty, during the writing of the book is most appreciated.

H.D.T.

1

Introduction

If plant pathologists and other agricultural scientists are to be effective in addressing the problems of food production in developing countries, the traditional farming systems in those countries must be thoroughly understood. This is essential so that researchers address appropriate problems in the context of farmers' systems, and so that efficient, proven techniques can be disseminated to other farmers. Traditional knowledge can be overvalued or romanticized, but that is better than despising or ignoring it. Far too many giant projects in developing countries costing huge sums have failed dismally and have caused serious ecological problems because they lacked sufficient understanding of traditional agriculture. Outside intervention is often misguided and irrelevant when it is undertaken without an understanding of traditional agriculture. Today there is serious concern about "modern agriculture" because it is so highly energy-intensive, its genetic base is narrow, and its goals of increasingly high yields and labor use efficiency lead to extensive monoculture and overproduction. Sometimes excessive erosion, pollution, and pesticide residues result. It is time to reexamine the potential for traditional agriculture to contribute to an improved, sustainable "modern agriculture."

Small farmers constitute a most important element in the agriculture of developing countries. Although figures vary somewhat, the following are typical. Data from the Food and Agricultural Organization (FAO) in 1970 indicated that holdings less than one hectare of land comprised 33%

Portions of this chapter are reprinted with permission from Thurston (1990), Plant Disease, Volume 74, copyright by the American Phytopathological Society.

of all holdings in developing countries. The mean size of agricultural holdings in those developing countries reporting to FAO was 6.6 hectares. According to the National Research Council (1982), "Half of the world's population is engaged in agriculture, the vast majority in the tropics and subtropics." Goodell (1984) wrote that small farmers till 65% of the world's arable land, and Todaro (1977) stated that 70% of the world's poor live in rural areas and engage primarily in subsistence agriculture. Most of the farmers in many developing countries are women. Poverty and socio-economic insecurity characterize the lives of a large sector of the rural population, and these problems are exacerbated in the vast group of small or traditional farmers who often have few resources beyond the labor of their families.

What is meant by traditional agriculture? The term traditional is usually associated with primitive agricultural systems or preindustrial peasant agriculture. Traditional farming usually is based on agriculture that has been practiced for many generations. Teri and Mohamed (1988) note that peasant production practices result from a long-term process of adjustment to the environment. Most small farmers in the developing world utilize agricultural practices that are to some degree "traditional", but many small farmers could not be characterized as traditional. The agricultural activities of traditional farmers are associated closely with their culture, as Schultz (1974) explains:

> Among primitive and peasant societies, cultural values and attitudes, beliefs and behavior patterns often play an equal or greater role than economic considerations when deciding whether to accept or not new production practices. Kinship obligations, peer group pressure, fatalistic beliefs, negative social sanctions regarding accumulations or surplus, individuality, caste differences and constraints and the perpetuations of common traditional values through family socialization all represent serious challenges to the foreign change agent.

Why Should Scientists Study Traditional Agriculture?

Anthropologists, archaeologists, ethnobotanists, and geographers -- and to a lesser degree ecologists, economists, and sociologists -- try to understand traditional agriculture. Unfortunately, plant pathologists and others in the so called "hard agricultural sciences" seldom take courses in these disciplines or read much of their literature, with the occasional exception of ecology and economics. Likewise, professionals in non-agricultural disciplines do not often read agricultural literature or take our courses in production sciences. Consequently each discipline develops a separate language, which is often all but unintelligible to outsiders.

Today the rhetoric in agriculture centers on "sustainability" (Edwards et al. 1990, Francis et al. 1990), LISA (low-input sustainable agriculture), and biotechnology. Although these terms are vague and all-encompassing, they strongly affect current funding and research directions. Some economists, strongly advocate continual growth in the world's economy, and others, more ecologically minded, believe that sustainable development should be the goal. Brown and Shaw (1982) stated: "In a world where the economy's environment support systems are deteriorating, supply-side economics -- with its overriding emphasis on production and near blind faith in market forces -- will lead to serious problems." Rapid economic growth rarely can be achieved without jeopardizing ecological sustainability. Some economists (Daly 1980) argue for a steady state economy rather than an expanding one. Schultz (1964), in his classic study *Transforming Traditional Agriculture*, suggests that a country dependent on traditional agriculture is inevitably poor. More recently Ruttan (1988) commented:

> Traditional agricultural systems that have met the test of sustainability have not been able to respond to modern rates of growth in demand for agricultural commodities. A meaningful definition of sustainability must include enhancement of agricultural productivity. At present the concept of sustainability is more adequate as a guide to research than to farming practice.

Do such conclusions by eminent economists suggest that nothing is to be gained by a study of traditional agriculture? I think not, and I doubt if such a conclusion is intended. If modern scientific agriculture is to have a role in the amelioration of world hunger and starvation, in part caused by population pressure resulting in environmental degradation, sustainable agricultural practices of traditional farmers in developing countries must be thoroughly understood and compared with alternative, new practices. If changes in traditional systems are necessary or needed, a thorough understanding of these systems is imperative as a first step before changes are initiated. Sustainable educational institutions of high quality with interest in and respect for traditional systems should be a high priority in future development funding. Traditional practices often provide effective and sustainable means of disease management. Traditional practices and cultivars (landraces) have had a profound effect on "modern agriculture," and most of our present practices and cultivars evolved from these ancient techniques and plant materials. Traditional systems and their disease management practices are in danger of being lost as agriculture modernizes; therefore, those practices should be studied carefully and conserved before they disappear.

Wilken (1974) suggests several additional reasons why the agricultural activities of traditional farmers are worthy of study. First, some traditional farming systems have excellent records in resource management and conservation. He suggests that those systems, which have lasted for thousands of years, surely justify serious study, although not all of the practices and strategies developed by traditional farmers are always successful. As Eckholm (1976) writes: "The littered ruins and barren landscapes left by dozens of former civilizations remind us that humans have been undercutting their own welfare for thousands of years." Perhaps we can learn from previous mistakes. In today's world a study of successful systems is especially important as petroleum, water, and other resources are becoming scarce.

Second, Wilken notes that although many traditional practices are labor-intensive, this aspect may be important and attractive to some societies having an abundance of labor and chronic unemployment. He writes that although traditional technology may be of little interest to scientists and Western businessmen, it represents the labor of millions of humans and the management of millions of hectares, and even small improvements would be significant for the world as a whole. For planners in developing countries, traditional methods have some advantages over modern agricultural techniques. For example, capital and technological skill requirements of traditional technologies are generally low, and adoptions often require little restructuring of traditional societies.

Finally, Wilken suggests that since modern agriculture has developed primarily in temperate regions the adoption of practices that are acceptable in these countries may have unexpected and undesirable impacts in developing countries, especially in the tropics.

Lack of Training in International Agriculture

To illustrate how lack of training in international agriculture can lead to errors in judgement when working with traditional systems in developing countries, I will use an example from my personal experience. In June 1954 I went to Colombia, South America as an assistant plant pathologist with the Rockefeller Foundation. My knowledge of the country was essentially zero. I had to look up Colombia's location in an atlas and knew not one word of Spanish.

Because of my lack of experience and training, I knew essentially nothing of the agriculture, customs, traditions, history, religion, or sociology of Colombia. I had seen Andean peasants only in picture books and had no inkling that thousands of years of agricultural trial-and-error, observation, and selection were behind what appeared to me to be haphazard or "primitive" farming systems.

I was hired by the Rockefeller Foundation to work in their agricultural program with the Colombian Ministry of Agriculture, specifically with potatoes. Fortunately, I did know something about potatoes, as I had received a M.S. degree in plant pathology from the University of Minnesota and had done my thesis on late blight of potatoes, a disease of worldwide importance. After a few months in Colombia, during which I had experienced a severe case of culture shock, but also had time to see how potatoes were grown and to travel a bit, I decided that almost everything the farmers were doing relative to growing potatoes was wrong. They planted whole tubers rather than cut seed as was done in Minnesota, they used very small tubers for seed (often 3-4 tubers per hill) rather than a single 30-40 gram seed piece of optimal size as was done back home, and they planted seed so there would be 50-60 cm between plants rather than the 20-30 cm recommended in Minnesota. Rows were 150 cm apart rather than the 90 cm row spacing Minnesota growers used.

The fungicides used for disease control were ineffective, herbicides were not used, storage procedures were inadequate, and so forth. Almost all cultural procedures were "*a mano*", i.e. done by hand. On steep hillsides (where I eventually realized that the vast majority of Colombian potatoes were grown) that was understandable, but in the level land of the Sabana de Bogotá, where our experiment station was located, I reasoned that large tractors and machinery such as that used in Minnesota were appropriate. Thus, I ordered a huge potato harvester, which simultaneously dug two rows of potatoes and put them directly into a truck. In retrospect, the machine was the most useless thing one could imagine for Colombian potato farmers and their conditions. Labor was less than $1.00 (US dollars) per day, and thus obtaining inexpensive labor for harvesting was not a major problem. The machine lasted barely two years before it broke down and became useless for lack of spare parts. By that time I had come to realize that perhaps it was not "appropriate" technology for Colombia.

Another order was for a 300-gallon, 14-row, John Bean potato sprayer. Insects and late blight of potatoes (caused by the fungus *Phytophthora infestans*) are serious problems in Colombia, and potatoes had to be sprayed frequently in order to obtain economic yields. The sprayer was useful for fungicides tests on our experiment station; we grew up to 100 hectares of potatoes on level ground, but it was not appropriate for most Colombian conditions. It took some time for me to realize that only a small percentage of the potatoes in Colombia could be sprayed with such a machine because of the steep slopes where most were grown. At that point we began using portable backpack sprayers for our fungicide tests, as most growers in the country used them, and the data we obtained using them was much more meaningful to Colombian growers than that obtained with a 300-gallon sprayer few of them could afford.

Almost all growers in the Andes of South America plant whole potato seed (tubers) rather than cut seed, which is commonly used in the United States. It is well known that cutting seed is an excellent way to spread pathogens (especially bacteria and viruses), but in the US we are able to use cut seed because of excellent seed certification programs and sound sanitation practices. Nevertheless, serious problems due to the use of cut seed still cause serious losses in the US. With my temperate zone mindset in 1954, I believed we should use cut seed as growers in Minnesota did, especially so we could use the tuber-unit method for reducing viruses. This is a method whereby a tuber is cut into four pieces and planted with a space between it and other tubers. This practice greatly facilitates field removal or roguing of virus-infected plants and in the 1950s was considered essential in the US to a good seed certification program. In 1955 the potato program of DIA (Division of Agricultural Research of the Colombian Ministry of Agriculture), in cooperation with the Caja Agraria (a semi-official agricultural bank in charge of seed production for DIA), began to increase supply of the improved variety Monserrate, which held great promise for potato culture in Colombia because of its high productivity, high degree of general resistance to *Phytophthora infestans*, adaptability, and other excellent agronomic characteristics (Estrada et al. 1959). Incidentally, Monserrate's high level of general resistance to *P. infestans* is still evident today (Parker 1989).

By 1959 a total of 700 tons of Monserrate seed was available for use by farmers. Almost all multiplication was accomplished using cut seed pieces, although customarily whole tubers were used in Colombia for planting. During the second growing season of 1959, about 30 hectares of Monserrate were planted by the Caja Agraria on the farm "Valmaria" near Bogota at an elevation of 2620 meters. This planting represented about 50% of the Monserrate seed available for the entire country for the coming season. At harvest time approximately 30% of the tubers were infected with *Pseudomonas solanacearum* (the bacterium that causes bacterial wilt of potatoes). The disease, although common on potatoes in many countries at lower elevations, had been reported only a few times at high elevations in Colombia. This loss was a severe blow to the potato program of DIA, since the infected seed from this farm had to be discarded or sold for human consumption. Similar seed in the hands of several growers who cut their seed, following DIA recommendations, produced fields with 100% infection by *P. solanacearum.*

As a result of these losses from bacterial wilt, growers and the Caja Agraria became convinced that Monserrate was highly susceptible to the disease and demand for seed declined drastically. In fact, the Caja Agraria almost terminated its national seed multiplication program. In subsequent years, when whole seed pieces were planted in the same fields, no detectable infection occurred. Our research program reverted entirely to using only whole seed, and subsequently we never had another problem with *P. solanacearum* on our station (Thurston 1963). We researchers finally came around to using a practice traditional farmers knew was practical for their conditions. Colombian farmers probably had discovered over the centuries that cut seed would not produce a crop. We scientists had to rediscover what the traditional peasant farmers of Colombia already knew. Many, although not all, of the practices of Colombian potato farmers had sound reasons for their existence, which we could not initially discern.

This example illustrates that because of lack of education or experience relative to traditional farmers and traditional agriculture in Colombia, my judgment on technology recommendations and appropriate areas of research in my first years there was initially poor. I spent a total of 11 years in Colombia and am proud of my association with DIA, ICA (Colombian Agricultural Institute), and the Rockefeller Foundation. In subsequent years, I believe I became useful and productive as regards the Colombian agricultural program, especially after I gained respect and appreciation for the knowledge of small farmers and the basic soundness of their farming systems.

Many projects intending to improve the lot of small farmers have failed due to a lack of understanding of how and why traditional agriculture works. I wish to emphasize that we in the temperate regions (our governments, universities, and private organizations) are still sending agricultural scientists to the tropics or into difficult, complicated environments with the same lack of training and experience I had initially. Often scientists who are sent have almost no understanding of or sensitivity to the agronomic and socioeconomic problems of the tropical regions; often they have the same mindset I originally had, i.e. that the only way to make progress is to do it like it was done back home. Not only the US, but most temperate countries of North America, Europe, and Asia are doing the same to some degree. Because of similar training, agricultural scientists trained in the leading agricultural universities of many developing countries encounter similar problems when they try to work with traditional farmers in their own country or other areas of the developing world.

Traditional Farmers' Knowledge

Traditional farmer knowledge is often impressively broad and comprehensive. A few examples can illustrate this. Conklin (1954) described the agricultural knowledge of the Hanunóo, a mountain tribe of Mindoro in the Philippines. On some aspects of agriculture their knowledge is amazingly wide, accurate, and practical. They distinguish 10 basic and 30 derivative soil and mineral categories and understand the suitability of each for various crops as well as the effects of erosion, exposure, and over-farming. They distinguish over 1,500 useful plant types, including 430 cultigens, and they discern minute differences in vegetative structure.

Mayan Indians in Mexico have their own comprehensive plant classification system. Berlin et al. (1974), describing the Mayan (Tzeltal) taxonomic system, state that "471 mutually exclusive generic taxa were established as legitimate Tzeltal plant groupings."

Bentley (1989) found that traditional farmers in Honduras, in addition to considerable general knowledge about plants, had an impressive knowledge of growth stages (phenology) of crops, especially maize and beans. Unfortunately, many farmers did not recognize plant diseases or pathogens. Bentley (1989) noted: "Traditional Central American peasant farmers know more about some aspects of the local agroecosystem than about others. In general farmers know more about plants, less about insects, and still less about plant pathology."

Much of the literature on traditional agriculture is anecdotal rather than experimental, much to the distress of scientists who believe that only information obtained by scientific methods is of real value. Also, traditional agriculture often includes a mixture of superstitious, religious, and magical beliefs (Casas Gaspar 1950). Some beliefs are of no obvious practical value, but others may constitute sound agricultural practices. Huapaya et al. (1982) interviewed Ayamara Indians near Lake Titicaca in Peru regarding their knowledge of plant disease management. They believe that plant diseases are caused by halos around the sun, certain phases of the moon, drought, hail, lightning, excessive humidity, fog, frost, dew, and the use of horse or cattle manure. The entrance into a field of animals in heat, pregnant or menstruating women, drunk men, or people or animals when dew is on the ground was also thought to cause disease. Indians dust their crops with ashes, spray them with fish water, place branches of *muña* (*Minthostachys* spp. -- a traditional insect repellent) between plants, and rogue diseased plants. To manage diseases they practice careful seed selection, crop rotation, and don't plant when the moon is full or the sun has a halo. People and animals are not permitted in fields when dew is on the ground. Several of the above practices would reduce disease incidence, but clearly their activities are a mixture of useful and useless practices.

Traditional Farmers' Practices for Managing Plant Diseases

Archeologists believe that humans began crop production perhaps 10,000 years ago. Some ancient societies developed sustainable agriculture practices that allowed them to produce food and fiber for thousands of years with few outside inputs; other traditional strategies were not so successful. Many of their successful practices have been forgotten or abandoned in developed countries, but are still used by many traditional, subsistence, or partially subsistence farmers in developing countries. Although considerable evidence shows that traditional farmers experiment and innovate (Chambers et al. 1990), most useful traditional methods of agriculture probably were developed empirically through millennia of trial and error, natural selection, and keen observation. These practices often conserve energy and maintain natural resources. Traditional farming systems, especially in the tropics, frequently resemble natural ecosystems. This, and their high level of diversity, appear to give them a high degree of stability, resilience, and efficiency. Teri and Mohamed (1988) state: "widespread plant disease epidemics in traditional agriculture are either rare, undocumented, unnoticed or all three." Traditional farmers are not always interested in the highest yields, but are concerned more with attaining stable, reliable yields. They minimize risks and seldom take chances that might lead to hunger, starvation, or loss of their land.

Most practices for disease management used by traditional farmers in developing countries are cultural practices. Yet little information is available in an easily accessible or understandable form on the cultural practices used in traditional systems. Palti's (1981) *Cultural Practices and Infectious Crop Diseases* is an excellent source of information on cultural practices for the management of plant diseases, but emphasizes primarily "modern" agriculture. Some practices of traditional farmers are these: altering of plant and crop architecture, biological control, burning, adjusting crop density or depth or time of planting, planting diverse crops, fallowing, flooding, mulching, multiple cropping, planting without tillage, using organic amendments, planting in raised beds, rotation, sanitation, manipulating shade, and tillage. Most, but not all, of these practices are sustainable in the long term. The disease resistance of traditional cultivars or landraces selected over millennia also is most important. Landraces are usually genetically diverse and in balance with their environment and endemic pathogens. They are dependable and stable in that, although not necessarily high yielding, they yield some harvest under all but the worst conditions. Pesticides are generally used in small amounts by traditional farmers, primarily because of their expense.

Sustainable Plant Disease Management Practices and Systems

A major question regarding traditional agricultural practices is: are they sustainable? Can a practice be continued for a long period of time without environmental degradation, serious reduction of crop productivity, and the addition of heavy fossil-fuel inputs? The information in Table 1.1 strongly suggests that most traditional practices are sustainable. However, note that some of the practices require high external inputs, and many practices have high labor requirements. The various practices in Table 1.1 have been characterized according to my own knowledge and perceptions, recognizing that exceptions to some classifications might be made.

In Table 1.2, a number of traditional agricultural systems are considered for their productivity (crop yield or income produced), sustainability (ability to maintain the system in existence over a very long period of time even when subjected to stress), stability (obtaining consistent and reliable yields in both the short and long run), and equitability (relative distribution of wealth in a society). Conway (1985, 1986) discussed and defined the above terms. My classifications are obviously subject to considerable discussion, and exceptions might be made to some the classifications. Table 1.2 illustrates considerations important to planning agricultural development.

Finally, systems should be analyzed for their yield per unit of land, capital, labor, and energy (other than human). Again, the classifications are generalizations I have made primarily for illustrative purposes. Table 1.3 gives insights into why traditional farmers make the decisions they do relative to their choice of farming systems.

Alternative agriculture, organic agriculture, ecological farming, and sustainable farming are among 16 terms used to describe a complex and interdisciplinary movement gaining rapid acceptance in recent decades, especially in the developed countries of the world (Merrill 1983).

APPLICATION OF CHEMICALS

2

Pesticides

Historically, many different natural insecticides have been used by traditional and indigenous farmers as numerous records attest (Schmutterer et al. 1987, Smith and Secoy 1975, Stoll 1987, Tait and Napopeth 1987, Yang and Tang 1988), but there is little on record about traditional usage of fungicides or other materials that control plant pathogens. Grainge and Ahmed (1988) cataloged 2,400 plant species controlling 800 pest species, primarily insects. Some of the references they listed described inhibition of fungi, bacteria, nematodes, and viruses. However, testing for inhibition was done primarily in the laboratory, not under farmer's conditions, using sap, oil, exudates, diffusions, extracts, or juice from plants.

In the eighth century B.C., Homer referred to "pest averting sulfur" (Keitt 1959), but the nature of the pest is unclear. Mason (1928) suggests that the earliest reference to a fungicide was Pliny's reference to Democritus (470 B.C.), who recommended that amurca of olives be sprinkled on plants to prevent blight. Unfortunately, the nature of the blight was not specified.

Today, if one drives through the beautiful olive-growing areas of Southern Spain, occasionally a foul odor intrudes. The unpleasant odor emanates from amurca (*alpechín* in Spanish), the liquid waste remaining after olives have been processed. Disposal areas for amurca are often found near olive-processing plants. Today in Spain and many other areas of the Mediterranean basin, amurca is considered a noxious substance causing serious environmental problems and polluting land, streams, and rivers. Its disposal is expensive and consumes considerable time, money, and labor. Legislation exists to prosecute those not properly disposing of amurca, but in practice this seldom occurs. Research is being carried out in Spain, Italy, Greece, and other olive-producing Mediterranean countries on a wide range of biological and physical methods for disposal of amurca. One possibility is using amurca as a fertilizer and soil conditioner. García-Rodriguez (1990) noted that the Spanish Moor Ibn al-Awam in the twelfth century (1988) and the Spaniard Alonso de Herrera in the sixteenth century (1988) recommended the use of diluted amurca as a fertilizer. The Roman Cato (234-149 B.C.) also recommended amurca for the improvement of soil fertility. Flouri et al. (1990) reported that when amurca was applied to agricultural land, tilth was improved, and the soil became suppressive to fungi such as *Phytophthora* and *Pythium* spp.

Amurca is mentioned frequently as a pesticide by ancient Roman and Spanish writers (Ainsworth 1981, Columela 1988, Alonso de Herrera 1988, Smith and Secoy 1975, Orlob 1973, White 1984). Amurca contains a small amount of oil, and vegetable oils are known to have some fungicidal effect against diseases (Calpouzos 1966). Martin and Salmon (1931) tested olive oil and found fungicidal action against powdery mildew. Amurca was almost a cure-all for the early Romans. Orlob (1973) suggests: "Oil dregs were the universal plant disease and pest remedy of antiquity, so to say a forerunner of today's fungicides and insecticides."

A rather long listing of the diverse recommendations for the use of amurca is given in Table 2.1. The listing is deliberately long to illustrate the variety of recommendations for the use of amurca over the centuries. Some repetition may occur, as some ancient authors did not cite their sources of information. There were six different kinds of amurca according to the Spaniard Alonso de Herrera (1988), writing in 1513: amurca from green olives, amurca from black olives, salted amurca, amurca without salt, crude or raw amurca, and cooked amurca. Crude amurca without salt was said to be the best for agricultural purposes, but several writers noted that it should be diluted with water as it was harmful to plants if too concentrated. Ancient writers recommended amurca for a wide range of plant problems, diseases, and insects. Oil dregs were even recommended for diseases of horses and sheep.

It would be interesting, and perhaps rewarding to determine whether the ancient uses of amurca (indicated in Table 2.1) were of any real value, and whether this substance, now causing such serious ecological problems, could be utilized to benefit, rather than harm, Mediterranean agriculture.

According to Mason (1928), Cato (234-149 B.C.) recommended fumigating trees for three days with the smoke from burning sulfur, amurca, and bitumum for control of the "vine fretter"). The nature of the "vine fretter" is unknown. Pliny (27-77 A.D.) suggested steeping wheat seed in wine or mixing bruised cypress leaves with wine to prevent powdery mildew.

TABLE 2.1. Recommendations for Amurca

Author	*Pest Problem*	*Recommendation*
Democritus (Greek - 470 B.C.) Cited by Pliny according to Mason (1928)	blight	sprinkle plants with amurca
Cato, Marcus Porcius (1934) (Roman 234-149 B.C.)	weak trees	spread or pour amurca around trees
	weevils and mice in grain	amurca around trees
	insects in grain storage	apply slime of amurca mixed with chaff to entire granary
	to keep caterpillar off vines	mix amurca with bitumen and sulfur and apply to trunk and under branches
	scab of sheep	apply amurca mixed with wine and lupine water
	moths in clothing	apply to wooden chest
	decay of furniture	rub amurca on furniture
	to keep figs from spoiling	place figs in earthenware vessel coated with amurca
Varro, Marcus Terentius (1934) (Roman 116-27 B.C.)	mice and worms in stored grain	treat stored grain and walls of granary with amurca

(Continues)

TABLE 2.1 (*Continued*)

Author	*Pest Problem*	*Recommendation*
Virgil (Roman 70-19 B.C.) Orlob (1973)	seed deterioration after planting	treat seeds with amurca before planting
Pliny (Caius Linus Segundus) (Roman 23-79 A.D.) Orlob (1973)	rust (Uredo) and cutworms	sprinkle cereals with amurca, then hoe
	diseases of seed and roots	treat seed with amurca and urine
Columela (1988) (Spanish Roman - 60 A.D.)	esca disease of grapes	split trunk of vine, insert soil dampened with amurca
	subterranean pests attacking grain	sprinkle soil with amurca
	faba bean weevils	water plants with amurca
	unthrifty fruit trees and olives	water with amurca plus wine
	ants, mice, and weevils in storage	water threshing floor with amurca
Bassos, Cassianos (Byzantine - 7th century A.D.) Orlob (1973)	weak olive trees	apply amurca
	excessive bleeding of grapevines in spring	make incision in vine and cover with amurca
Palladius (Roman - 4th or 5th century A.D.) Orlob (1973)	unhealthy olives	pour amurca and urine on roots
	unhealthy peach tree	treat with amurca

(*Continues*)

TABLE 2.1 (*Continued*)

Author	*Pest Problem*	*Recommendation*
	sick plum tree	treat with brine, water, and amurca
Ibn al-Awam (1988) (Moor, Seville, Spain, 12th century)	grapes produce fruit that shrivels on vine	mix amurca with groundup nuts and vinegar, apply for 20 days
	excessive bleeding of grapevines	make incision in stem and cover with amurca mixed with ground pistachios and barley flour
	grape insects	fumigate with amurca
	insects (general)	spray with amurca
	grasshoppers and worms	spray plants with amurca mixed with bile
	stored grain	spray with amurca
	worms in vegetables	spray with amurca and bull urine
Alonso de Herrera (1988) (Spanish 1513 A.D.)	storage insects in grain (worms, weevils)	mix amurca in clay to plaster walls and floor of storage
	dry grape vines	apply amurca mixed in clay
	ants	apply amurca mixed in clay

(*Continues*)

TABLE 2.1 (*Continued*)

Author	*Pest Problem*	*Recommendation*
	grape insects	smoke vines with burning amurca
	figs falling prematurely	soil application
	insects (general)	soil application
	rats	soil application
	moles	soil application
	bugs in beds	soak wood used to make bed in amurca
	worms in storage jars	soak jars in amurca
	worms in planting seed	soak seed in amurca
	snails in chick peas	sprinkle plants with amurca

Chiu and Chang (1982) mentioned that the use of sulfur as a pesticide was recorded in a Chinese book published in 1313, and that hot water treatment of cotton seed before sowing was recommended in 1765.

According to Orlob (1973), Ibn al-Awam's book "*Libro de Agricultura*" (1988) was the most comprehensive treatise produced by medieval agriculture. An entire chapter in his book was concerned with the control of plant diseases. Ibn al-Awam, also known as Abu Zacaria Iahia and Aben Mohamed Ben Ahmed Ebn el Awam, was an Arab writer of twelfth century Southern Spain. He frequently mentioned ashes as a remedy for several diseases and insect problems. Ashes were also recommended for disease control in ancient India according to Raychaudhuri (1964). Common recommendations included the use of ghee, fat, cow's horn, black pepper, mustard, honey, milk, urine, and various dungs (Raychaudhuri 1964). A compilation of the superstitions, ceremonies, magic, and religious practices and traditions of ancient farmers regarding disease management would fill many books. Some of today's scientists scoff at ancient practices and consider them absurd; perhaps many of them were. Orlob (1973) in his study of ancient and medieval plant pathology gives several examples. For example, he cites the Roman writer Columella as suggesting that to avert "rust" a skinless donkey head be placed at the edge of the field. The Romans had a god of "rust", Robigus, to whom they prayed for protection from cereal rust. A religious festival involving Robigus was held on the outskirts of Rome on April 25 in order to obtain a healthy cereal crop (Ordish 1976). Orlob (1973) also cites the following translation from the Greek text of the "Geoponica," written about 350-400 A.D.:

> When you observe "rust" being formed in the air, burn at once the left horn of an ox together with cuttlefish. Surround the field with plenty of smoke so that the wind may carry the smoke towards the rust. For the smoke scatters the air that causes the damage."

In the fourteenth century a Spanish priest excommunicated worms consuming the harvest (Casas Gaspar 1950). Poma de Ayala (1987) describes an Inca festival of the moon in September, which included the removal of pestilences and diseases from the empire. Armed as if they were going to war, men ran through the streets carrying fire and shouting "Depart, diseases and pestilences from this people. Leave us." Although most of the ceremonies were directed at human diseases, a disease of maize (*sara oncuy* in Quechua, the Inca language) was included. Before we pass too harsh a judgement on such practices, we might remind ourselves that farmers in the US have often hired American Indians to do "rain dances" when drought has become severe. Almost all peoples appeal to their deities for help in times of agricultural stress.

Although traditional pest management practices often included superstitious practices, many ancient practices were highly effective. Most practices were developed empirically through millennia of trial and error, natural selection, and observation.

Pesticide Use in Traditional Systems

Overuse of modern pesticides in traditional systems is common today (Figure 1.1). Although traditional farmers may have considerable knowledge regarding their agroecosystem, their knowledge seldom includes information regarding modern pesticides, and usually they must rely on sources outside of their traditional culture for information.

Overuse and misuse of chemical pesticides is often tragic, as the following dramatically illustrates. In Iraq from 1971 to 1972 an estimated 6,000 people died and 100,000 may have been injured by eating imported wheat seed treated with mercurial pesticides. Two years of severe droughts had caused a famine, and desperate people ate imported wheat seed to keep from starving. Although the seed had been dyed pink to show it was poisonous, many peasants obviously did not realize that the seed was poisonous (Hughes 1973). Hundreds of additional examples of misuse of pesticides in developing countries could be cited. Ewell and Merrill-Sands (1987) found that Mayan farmers who entered the market economy were spraying vegetables every 4-5 days with "cocktail" mixtures of pesticides. There are many books and articles written relative to the abuse of pesticides in traditional societies (Bentley 1989 , Bull 1982, Chapin and Wasserstrom 1981, Goodell 1984, Gunn and Stevens 1976, Repetto 1985, Wasilewski 1987).

It is not always essential to use pesticides in agriculture. For example, in Guangdong province, Southern China, a dike-pond farming system exists, which covers 800 square kilometers and sustains a human population of over a million. This intensive system includes aquaculture, animals, and multiple cropping of a variety of food, feed, and cash crops. Up to 12 crops of leafy vegetables are harvested annually from the dikes. According to G. L. Chan (personal communication), no pesticides are used in this highly productive system.

Today, despite numerous instances of overuse, the quantity of pesticides used by most traditional farmers is still very small. The high cost of pesticides seriously limits their use in developing countries, as few farmers can afford to use them because of their high cost. However, their expectations for pesticides are often unrealistically high. For example, Rosado and Garcia (1986) interviewed fifty-nine farmers in Tabasco, Mexico regarding to their management methods for the fungus disease web blight of beans (*Thanatephorus cucumeris*). Disappointingly, although they used several cultural methods of management, all of the farmers interviewed said they were expecting a chemical solution to the problem.

Traditional Fungicides

Ashes are frequently used by traditional farmers as a fungicide. Ash also may remedy minor element deficiencies. Wilken (1987) mentions that in Mexico ash from kitchen fires is saved, so it can be used later to dust over plants to prevent fungus infection. No information was given on its effectiveness. Huapaya et al. (1982) also cite dusting crops with ashes for plant disease management. Traditional farmers in Mali and Senegal mix ash with threshed millet before storage, and little damage due to storage fungi was reported (National Research Council 1978). The ash was also scattered on the floor of the millet storages, rubbed into the walls, and mixed with the threshed grain. Zehrer (1986) wrote that farmers in West Africa mixed ashes with beans and Bambara groundnuts before storage. Traditional farmers in Ghana coat yam setts with ash before planting (Adesiyan and Adeniji 1976). Many writers have recorded the use of ashes to prevent food deterioration in storage (Ibn al-Awam 1988, Jurion and Henry 1969, Orlob 1973, Sagnia 1989, Upawansa 1989, Varro 1934, Zehrer 1986).

Summary

Historically, chemical pesticide use by traditional farmers has been relatively minor. Today, chemical pesticide use is a major plant disease management strategy worldwide, and traditional farmers often have unrealistically high perceptions of the value of such pesticides. Pimentel (1981) cited estimates that in 1985 over 2.7 billion pounds of pesticides would be used worldwide. Unfortunately, although the short term effects of pesticides may be dramatic and highly effective, the long term effects may be disastrous. There are ever increasing numbers of pests, including insects and fungi, that are resistant to pesticides. Toxic effects on humans, wildlife, and the environment are common if pesticides are not properly used.

The extensive traditional use of ashes as a fungicide would suggest that research on its effectiveness would be a worthwhile objective for governments and development agencies. Traditional farmers may have an in-depth knowledge of their environment and their farming systems, but they almost never have knowledge of the chemistry, toxicology, or even proper application of today's pesticides. Farmers must usually rely on outside sources for such knowledge, and the most common source is often the pesticide salesperson. Perhaps the high cost of pesticides is a blessing in disguise, as most traditional farmers seldom can afford to use large quantities of pesticides because of their high cost. There are situations where the use of pesticides may be a safe and efficient practice for plant disease management, but generally recommendations of pesticide use for traditional farmers should be made only with great caution, after thorough study and evaluation of alternative management practices. Because of the high cost and toxicity of chemicals, an objective of plant disease management strategies in developing countries should be to reduce or to eliminate chemical pesticide use.

BIOLOGICAL CONTROL

3

Biological Control

The science of biological control is only a century old and has primarily concentrated on insects until recently (Nelson 1989). However, today there are several excellent and comprehensive books on biological control of plant pathogens (Baker and Cook 1974, Cook and Baker 1983, Papavizas 1981). With regard to control of plant pathogens Cook and Baker (1983) define biological control as "the reduction of the amount of inoculum or disease-producing activity of a pathogen accomplished by or through one or more organisms other than man." The destruction or reduction of one organism by another is common in nature. Biological control is used in a variety of ways to manage plant pathogens.

Traditional farmers have used biological control through their development of suppressive soils and the use of antagonistic plants. The addition of large amounts of organic matter to soils by Chinese farmers, which often results in suppressive soils, is probably one of the oldest biological control practices. Youtai (1987) writes that the value of adding manure to field soils was established in China before the fifth century B.C. An example of biological control of insects by traditional farmers is given by Huang and Yang (1987), who state that the yellow citrus ant has been used by peasants in China for 1,700 years to protect citrus fruit against insect pests. Farmers in China are still culturing the ants for use in citrus. Classic modern examples of the success of biological control, such as the use of insect parasites and predators in the Cañete Valley of Peru to control cotton insects, are given by Boza (1972), Ewell et al. (1990), and Smith and Reynolds (1972).

Most cultural methods of control have a direct effect on the efficiency of biological control. The relationship between cultural control and biological control should always be kept in mind when planning or analyzing plant disease management strategies.

Suppressive Soils

Various definitions are given for suppressive soils. Cook and Baker (1983) define suppressive soils as soils in which disease development is suppressed, although pathogens are introduced in the presence of a susceptible host. There is an extensive literature on suppressive soils. Baker and Cook (1974), Cook and Baker (1983), Hornby (1983), and Schneider (1982) have comprehensive reviews of the subject. Suppressive soils are known for a large number of pathogens, primarily fungi and actinomycetes. Palti (1981) gives 13 examples of soils that suppress fungal pathogens. Hornby (1983) discusses the various mechanisms that have been suggested for soil suppressiveness.

Shipton (1977) identifies two patterns of disease occurrence during monoculture. In the first or irreversible pattern, the disease incidence tends to become constant in some soil pathogen/host combinations. In the other reversible pattern disease develops, but then tends to decline over an extended period of time as suppressive soils are developed. Take-all of wheat (caused by *Gaeumannomyces graminis* var. *tritici)* is commonly used as a classic example of disease decline. Worldwide, *G. graminis*, the causal agent of the take-all disease of cereals and grasses, is a major limiting factor to cereal and grass production to which no effective chemical control or genetic resistance has been found. According to Baker and Cook (1974) take-all severity generally increases for 2-4 years under wheat monoculture and then declines in the following years of wheat monoculture. The biology and management of take-all has been extensively reviewed (Asher and Shipton 1981).

Cook and Baker (1983) noted that about 80% of the People's Republic of China's fertilizer requirements were met with organic sources such as composted crop residues, green manure, human waste, and livestock manure. Over 100 t/ha of compost were often used annually in Chinese agriculture. Pathogens are often killed by heat generated during the composting process, and many diseases are suppressed by the use of composts (Hoitink and Fahy 1986). Cook and Baker (1983) further stated:

> Perhaps the best large-scale demonstration of effective biological control by cultural practices is the widespread multiple-cropping organic system used in the People's Republic of China. The agriculture of that country, which feeds nearly one fourth of the earth's population, clearly demonstrates that farming can be both intensive and sustainable and, if stabilized for years or perhaps for centuries, can provide a biological balance and disease suppression similar in effect to the disease suppression that can occur with prolonged monoculture of some crops.

Cook and Baker (1983) later observed:

> A nation such as China is less able than many western nations to afford modern synthetic pesticides, except in specialized situations, and hence they continue to use practices that provide biological control. Any shift within such a country towards a western-type agriculture with more intensive use of chemical pesticides should be made only if it improves or at least does not upset the existing biological controls.

I shall discuss the use of organic amendments in greater detail in Chapter 13.

Lumsden et al. (1987) studied *chinampa* soils near Mexico City relative to disease. *Chinampas* constitute a raised field system used for centuries by the Aztecs in the Valley of Mexico. The system is perhaps better known as the "floating gardens" found near Mexico City, Mexico. This ancient sustainable system is described in greater detail in chapter 14. Relative levels of damping-off disease caused by *Pythium* spp. on seedlings grown in soils from the *chinampas* were compared with levels in soils from modern systems of cultivation near Chapingo, Mexico, and it was found that disease levels were lower in the *chinampa* soils (Lumsden et al. 1987). When they introduced inoculum of *Pythium aphanidermatum,* the fungus was suppressed by *chinampa* soils. From their studies Lumsden et al. (1987) concluded:

> In the chinampa agroecosystem, apparently a dynamic biological equilibrium exists in which intense management, especially of copious quantities of organic matter, maintains an elevated supply of organic nutrients and calcium, potassium and other mineral nutrients which stimulate biological activity in the soil. The elevated biological activity, especially of known antagonists such as *Trichoderma* spp., *Pseudomonas* spp., and *Fusarium* spp., can suppress the activity of *P. aphanidermatum,* other *Pythium* spp. and perhaps other soilborne plant pathogens.

More recently Zuckerman et al. (1989), in a cooperative study between scientists from Mexico and the United States, also studied disease suppression in Mexican *chinampa* soils, but of plant parasitic nematodes rather than fungi. The authors pointed out that the high organic content of the soil is probably responsible in part for the relatively few nematodes in *chinampa* soils, but they also found nine organisms with antinematodal activity. Their results were summarized as follows:

> Soil from the Chinampa agricultural system in the Valley of Mexico suppressed damage by plant-parasitic nematodes to tomatoes and beans in greenhouse and growth chamber trials. Sterilization of the chinampa soils resulted in a loss of the suppressive effect, thereby indicating that one or more biotic factors were responsible for the low incidence of nematode damage. Nine organisms were isolated from chinampa soil which showed antinematodal properties in culture. Naturally occurring populations of plant-parasitic nematodes were of lower incidence in chinampa soils than in Chapingo soil.

Castillo (1985), Muller and Gooch (1982), Rodriguez-Kabana (1986), Rodriguez-Kabana and Morgan-Jones (1987, 1988) and Sayre (1971) have noted the usefulness of organic amendments in promoting suppression of nematodes in soil. When chitins (as crustacean, fish, and other animal wastes) are added to soil there is an increased parasitism of nematode eggs by fungi (Rodriguez-Kabana 1986, Rodriguez-Kabana and Morgan-Jones (1988). Additions of chitin appear to increase the number of chitin-decomposing organisms and chitinase activity in the soil, thus increasing egg parasitism.

Popal grass (*Thalia geniculata*) grows in frequently flooded, swampy areas of Tabasco, Mexico. Traditional farmers use a highly productive system (*popal* system) in which a cultivar of maize (*Marceño*) is planted in deep holes (10-20 cm deep) in fields covered with popal grass, which has previously been cut and allowed to dry. Immediately after the maize emerges the grass is burned, but the maize survives and grows to give yields of 4-5 t/ha compared to average yields of 1.2 t/ha in other maize plantings in Tabasco (Garcia Espinosa 1987). Garcia-Espinosa (1980b) inoculated Tabasco soil grown to maize (and having a declining maize production due to soilborne pathogens) and *popal* soil (which has a 30% organic matter content) and found evidence of suppression of *Pythium* spp. in the *popal* soils. Lumsden et al. (1981) found that *popal* soils were suppressive to *Pythium aphanidermatum, Sclerotinia rolfsii,* and *Rhizoctonia solani.*

Antagonistic Plants and Trap Crops

Cook and Baker (1983) define antagonists as "biological agents with the potential to interfere with the life processes of plant pathogens." Antagonists include all types of microorganisms, including seed plants. Some traditional farmers promote biological control by the use of copious amounts of organic matter and thus encourage the activities of microbial antagonists. Organic amendments generally produce enhanced competition among microorganisms for nitrogen, carbon, or both, and this may result in fewer soil pathogen problems(Cook and Baker 1983).

A number of plants contain chemical compounds that are antagonistic to various plant pathogens. Trap crops or trap plants are also considered to be antagonists and a form of biological control by Cook and Baker (1983). Nematodes enter the roots of trap crops, but fail to develop and subsequently die. For example, root-knot nematodes (*Meloidogyne* spp.) enter the roots of *Crotalaria spectablis*, but do not survive (Cook and Baker 1983). *C. spectabilis* is also often used as a cover crop and subsequently turned under as a green manure (Birchfield and Bistline 1956, Sasser 1971). Yoshii and Varon (1977) found both *Tagetes minuta* and *C. spectablis* reduced soil nematode populations and contributed to increased soybean yields in Colombia. In coastal Peru, where there are high populations of root-knot nematodes in the soil, farmers know that beans planted after marigolds suffer less from root-knot damage. Fields of marigolds are grown in the area, and the flowers are used as a additive to color egg yolks (Personal communication -- Barbara Mullin).

Marigolds (*Tagetes* spp.) are a highly regarded flower in many Mexican traditional societies and are commonly planted in and around maize fields. The genus *Tagetes* is native to the Americas. Some 33 species of marigolds are known, and some, such as *T. patula* and *T. erecta,* also act as trap crops for nematodes (Belcher and Hussey 1977, Cook and Baker 1983). *Tagetes* spp. produce terthienyls toxic to nematodes and also to some fungi according to Baker and Cook (1974). The book by Suatmadjii (1969) reviews the literature on the effect of *Tagetes* spp. on nematodes.

Friar De Sahagun (1969) arrived in Mexico in 1529. He worked there for 61 years, learned Nahuatl (the Aztec language), interviewed many Aztec informers, and had drawings made of most aspects of Aztec life. His *Historia General de las Cosas de Nueva España* is the most complete description of the Aztecs written. De Sahagun (1969) and Kaplan (1960) refer to marigolds in the sixteenth century as volunteers and also as being planted in gardens in Mexico. Kaplan (1960) also noted that in Mexico marigolds are commonly found between the rows in maize fields. Marigolds are not treated as weeds by traditional Mexican farmers, but rather are allowed to remain when other weeds are removed. De Sahagun referred to them by their Nahuatl name (*cempoalxochitl*). In Mexico today, they are often called the "*Flor de Muerto*" (flower of the dead) and are associated with traditional ceremonies for the dead. De Sahagun (1969), writing in the sixteenth century, described an Aztec ceremony at which human sacrifices were made, and stated that all the people watching the ceremony had yellow flowers -- marigolds -- in their hands. Marigolds are used in Mexico today to decorate household altars. Little information was found on whether marigolds have any effect on the management of nematodes or other pathogens in Mexico. Another plant found to have nematicidal properties in Mexico is *Chenopodium ambrosioides* (Garcia 1980a).

Crops highly susceptible to nematodes have also been used as trap crops (Godfrey and Hoshino 1934, Whitehead 1977). Plants are allowed to grow long enough for the nematodes to enter the roots, and then the crop is destroyed and nematode populations are reduced. Timing is obviously most important with this method of management.

Brodie (1982) and La Mondia and Brodie (1986) have used resistant potato cultivars experimentally as trap crops. Nematode populations are reduced considerably using this method, as nematodes enter the roots of resistant cultivars, but subsequently are unable to reproduce.

Mayer (1979) and Brush (1977) indicate that traditional Andean farmers not only leave their fields fallow, but use other crops in their rotations. A crop used in rotation with potatoes by ancient Peruvian farmers was mashua (*Tropaeolum tuberosum*). Johns and Towers (1981) and Johns et al. (1982) found that mashua contains isothiocyanates, which are nematicidal compounds. (Johns et al. 1982) reported that in Cuyo-cuyo, Peru, local farmers plant mashua interspersed among other tuber crops because they consider it resistant to pathogens. Thus, mashua may have acted as a trap crop in the Inca rotations.

Summary

Cook and Baker (1983) cite Chinese agriculture as an example of "the extensive adoption of biological control measures." The measures referred to consist primarily of the extensive use of organic amendments. Historically, many sustainable agricultural systems incorporated large quantities of organic matter into soil. This incorporation generally resulted in less soilborne disease, in addition to other important agronomic benefits, and the practice should be recommended whenever feasible. The value of organic amendments will be further discussed in Chapter 13.

Biological control of plant pathogens is still an emerging area of plant pathology. In 1983 Cook and Baker wrote that there were only six examples of biological control of plant pathogens and only two of these were used commercially. If that was the state of biological control in modern agriculture in 1983, the scarcity of sound information on the use of biological control by traditional farmers is not surprising.

Few examples of the use of antagonistic plants (trap crops and trap plants) by traditional farmers were found in the literature. Their use by traditional farmers is probably far more widespread than the literature indicates and should be considered for management of nematodes and other soilborne pathogens.

CULTURAL PRACTICES

Planting

4

Adjusting Density and Spacing

The proper distance of planting has been a concern of farmers for centuries. Columela (1988), a Roman who lived in Spain in about 50 A.D., discussed at length factors affecting the quantity of cereal seed to be planted in a given field, and their effect on crop density. He mentioned soil fertility, soil type, soil tilth, the slope of a field, different climatic conditions, time of year, moisture, and the presence of other crops in the field. Ibn al-Awam (1988), writing in Islamic Spain during the twelfth century, gave the recommended distance of planting for dozens of crops grown at that time.

Plant pathologists have made relatively few studies of the effect of plant or crop density on disease, although it is recognized that dense populations usually contribute to disease epidemics. Burdon (1978) cited several studies showing that the epidemic rates of disease increase with increasing crop density. Antonovics and Levin (1982) and Burdon and Chilvers (1982) reviewed the literature on host density as a factor influencing plant disease. Most (57%) of the 69 references Burdon and Chilvers reviewed showed a positive correlation between host density and disease incidence; however, 35% gave a negative correlation. Those references showing a positive correlation primarily concerned fungal diseases, while over half of those with negative correlations related to viral diseases. Thresh (1982) has discussed crop densities (stands) as they affect virus transmission. Cowling (1978) noted that in dense plant stands the distance pathogens or their vectors have to travel is reduced, splash dispersal of inoculum becomes easier, and wounding during cultivation may increase. Leaves may touch and root contacts increase. In closely packed plants the microenvironment changes; temperatures become more uniform, relative humidity increases, and leaves stay wet longer after rain or dew.

Manipulations of Plant Density

Plant density is affected by many practices such as spacing (between plants or between rows), pruning, thinning, fertilization, water management, staking, trellising, and harvesting plants or plant parts (Palti 1981). Host density is further manipulated by intercropping, and one of the reasons for the common use of intercropping by traditional farmers may be its important role in disease management.

Increased Disease in Dense Stands

Most commonly, dense stands increase disease incidence and severity. Closer spacing causes the crop canopy to close more rapidly, producing a cooler, more humid microenvironment, which is generally more conducive to disease development. Free water is necessary for the germination and penetration of many fungal spores, and dense stands allow free water to remain on plant parts for longer periods than in widely spaced stands. Allen (1977, IITA 1976) noted that the amount of shade varied with the spacing of maize interplanted with cowpeas, and that powdery mildew of cowpeas (*Erysiphe polygoni*) increased with increasing shade. Campbell (1949) reported less grey mold of beans (*Botrytis cinerea*) with wider spacing. Burke (1964) found less fusarium root rot of beans (*Fusarium oxysporum* f. sp. *phaseoli*) when plants were widely spaced. Experiments in IRRI (1979) showed that closer spacing of maize in intercropping experiments with mung bean increased powdery mildew (*E. polygoni*). Amin and Katyal (1979) reported that as the seeding rate in rice was increased, the incidence of rice blast (*Pyricularia oryzae*) increased. Kozaka (1965) reported that close spacing tended to increase rice blast (*Pyricularia oryzae*) in Japan, but that in non-fertile soil this was not always the case.

In the 1950s, due to increased demand for seedlings, nurserymen expanded their nursery size and increased seedling density in their nursery beds. Crowding of seedlings resulted in serious epidemics caused by species of *Pythium*, *Fusarium*, *Rhizoctonia*, and *Cylindrocladium* (Cowling 1978).

Experimentally, wider spacing was reported to reduce disease caused by several pathogens (Berger 1975, Steadman et al. 1973). Palti (1981) gave examples of increased disease in dense crop plantings. Foliar pathogens that are favored by high moisture, pathogens whose inoculum is disseminated by splashing soil, and soilborne fungi and damping-off organisms are more serious in dense crop stands. Palti noted that transmission of viruses, and plant-to-plant infection by other pathogens, were increased by plant contact. Thresh (1982) wrote that dense stands facilitated virus transmission by pollen and active arthropod vectors, in addition to transmission by vectors such as fungi, nematodes, and wingless insects, which do not move long distances.

The work of Autrique and Potts (1987) showed that intercropping potatoes with maize and beans reduced the incidence and rate of bacterial wilt development (*Pseudomonas solanacearum*) in potatoes. The reduction was affected by increased distances between individual potato plants and the presence between potato roots of roots of other crop species. They conclude that "the use of low plant densities and crop association, as presently practiced by many farmers in developing countries, is an efficient and complementary means of aiding in the control of the disease."

South American leaf blight of rubber, (*Microcyclus ulei),* has been the major problem of rubber production in the Americas (Thurston 1984). In the native habitat of Hevea rubber trees in the Amazon jungle there are only a few Hevea trees per hectare (Imle 1978). Many of the native rubber trees were low-yielding and somewhat tolerant of the disease; thus, the damage done by *M. ulei* was not as serious as the damage to selected high-yielding trees grown in a monoculture. In addition each rubber tree in the forest was screened from the other rubber trees by the foliage of trees of other genera, which served as barriers to windborne spores. The disease did little damage during the time when rubber was primarily collected from the forest (Langford 1945). However, when higher yielding Hevea rubber trees were grown as a monoculture in large plantations, they had more exposure to attacks by *M. ulei.* The disease became a serious problem and destroyed thousand of hectares of rubber in Latin America. The natural wide spacing of rubber trees in the Amazon forests gave some protection from *M. ulei,* which devastated monocultures of Hevea planted in the Amazon such as those started by the Ford motor company in Brazil in 1928 (Imle 1978, Thurston 1984).

Decreased Disease in Dense Stands

Less commonly, dense stands may reduce disease incidence and increase yields. An example is given by Allen (1983) with the groundnut (peanut) rosette disease in Africa. A'Brook (1964 and 1968) and Davies (1976) found that early planting and close spacing increased yield and reduced the incidence of the virus. It was theorized that the aphid vector (*Aphis craccivora*) was attracted more strongly to broken ground cover than continuous ground cover. For example, weeding actually increased the disease incidence (Hayes 1932).

Have and Kaufmann (1972) found that wider spaced rice plants had more severe bacterial blight (*Xanthomonas campestris* pv. *oryzae*). They postulated that at the later stages of growth the wider spaced plants had a higher nitrogen content and that predisposed them to a higher level of disease. Even under conditions of severe disease, the highest yields were obtained in the closer spacings, as positive effects of high nitrogen levels tended to mask differences in disease due to spacing.

The density of crop cover, especially that of tropical trees, has important effects on disease incidence. Waller (cited by Palti 1981) described the effect of the density of plant cover on tropical plant diseases as follows:

> In tropical plantation crops, density of plant cover may have a twofold effect. In the rainy season, when rain runs down limbs and trunks, wet soil and foliage will take longer to dry under dense cover, and prolonged periods will favour many diseases, such as the coffee berry disease (*Colletotrichum coffeanum*). Conversely, in seasons poor in rain but rich in dew, dense plant cover will shield lower organs from dew formation, and will thus reduce the proportion of shoot growth in danger of attack by pathogens requiring films of water for their development.

Practices of Traditional Farmers for Manipulating Crop Density

Friar Francisco Javier Clavigero (1974), who died in 1787, described the planting of maize with sharpened planting sticks by Aztec Indians. Their distance of planting varied according to the quality of the soil. He also said their plantings were so straight that they appeared to have used a cord, and the distance between each plant was so equal that it appeared to have been measured. It is obvious that the distances between plants and between rows in different types of soil were of importance to Aztec farmers and that they were experts in manipulations of plant spacing.

In Indonesia most cultivars of maize are susceptible to downy mildew (*Peronosclerospora maydis*), but indigenous farmers usually space rows at wide intervals and often plant rice between the rows of maize. This spacing facilitates air circulation and rapid drying, so that the maize plants dry rapidly in the morning, thus reducing the number of hours favorable for fungal infection. Sastrawinata (1976) noted that the density of planting maize significantly affected mildew levels, provided the infestation was not extreme. Harwood (1979) mentioned that when maize is grown in Southeast Asia in rows two to three meters apart and intercropped with other crops, such as mung bean, rice, peanuts, or soybeans, the maize has little downy mildew. Poltthanee and Marten

(1986) describe an intercropping system in Thailand where green corn is grown on the shoulders of beds and rice is grown in the furrows inbetween. Davis (1988) described the system used by settlers in the less densely settled outer islands of Indonesia:

> Today, the standard farming system used by migrants involves planting corn at wide intervals with the first rain, interplanting rice as the corn matures, and establishing manioc in or around the field as the rainy season advances.

The traditional spacing practices described above are useful in managing *Peronosclerospora maydis*.

Hirst and Stedman (1960) found that as the density of potato foliage increased in the field, the microclimate became more favorable for *Phytophthora infestans* (late blight). The conditions within a dense canopy of potato foliage are usually ideal for infections by *P. infestans* in comparison to those found in sparse, easily aerated foliage. In the Andes of South America rows are often planted 150 or 200 cm apart rather than the 90 cm more common in the US. Although many South American *Solanum andigenum* cultivars grown in the Andes produce large vines, the wider spacing of rows in the Andes may also reduce problems with late blight.

Whether late blight of potatoes caused by *P. infestans* occurred in South America before 1845 has been debated for some time. Most authorities on *P. infestans* state that the fungus did not occur in South America before the middle 1800s, but rather originated in Mexico where the perfect stage is found in mountainous areas on wild Solanum species. De Acosta (1987), who wrote in the sixteenth century, and who had lived and traveled extensively in the Andes of South America, observed: "Finally, these roots are the bread of that land, and when the year is good for them, they are content, because very many years they are blighted and frosted in the same land; such is the cold and harshness of that region." It is not clear whether De Acosta was describing a blight caused by *P. infestans*, frost damage, or both, but since potatoes originating in the Andes exhibit a wide range of general resistance to the fungus (Thurston 1971), it is reasonable to conclude that De Acosta was describing damage caused by *P. infestans* .

Trutmann et al. (in press) interviewed traditional farmers in the highlands of East Africa who stated that bean plants in damp and fertile conditions should not touch each other. Planting density was manipulated to prevent touching. In drier, less fertile conditions plant density was not decreased, as yields could be reduced. Rate of sowing was also altered depending on soil fertility, weed pressure, and seed viability. Manipulations also included training climbing beans on stakes, and farmers stated that staked plants should not touch each other. The investigators noted that avoiding plant contact and reducing humidity by regulation of plant density reduced disease incidence.

Ignacio de Asso (1947) wrote in Spain in 1798 that rust of snap beans (*Uromyces appendiculatus*) could be prevented by staking the plants to give them better ventilation. Without staking the beans were lost to rust.

Alterations of crop architecture are commonly made by traditional farmers. Such practices affect the crop microclimate and may significantly reduce the incidence of some diseases. Staking and pruning were also used by traditional farmers to alter the architecture, and thus the crop density, of beans in East Africa (Trutmann et al. -- in press). Plant architecture is important in the management of web blight of beans (*Thanatephorus cucumeris*). The upper foliage of some bean cultivars usually escapes infection, since inoculum contained in splashing rain cannot reach it. Schwartz and Galvez (1980) suggested that upright plant architecture, an open canopy, and wide plant spacing all contribute to the management of web blight. The movement of air within a crop canopy affects the dispersal of plant pathogens and their insect vectors. Air movement also affects temperature, humidity, and dew deposition. Farmers in Tabasco, Mexico increased distance of planting between bean plants for better management of web blight of beans (*T. cucumeris*) in an area where yield losses of up to 95% of bean production due to web blight have been recorded (Rosado May and Garcia 1986).

Adequate aeration is important in the prevention of many diseases, and good crop ventilation is enhanced by sowing or planting in rows parallel to the direction of prevailing winds. Planting on steep gradients, especially in mountainous regions, may also affect wind velocity and thus crop aeration. Thinned foliage has better ventilation and light penetration and reduced humidity, and allows quicker drying; these generally lead to less disease.

One of the major problems of the high-yielding varieties (HYVs) of rice in Asia is sheath blight caused by *Thanatephorus sasakii*. Ou (1972) wrote that the disease is especially destructive under conditions of high humidity and high temperature. The density of plant stands greatly affects humidity, and the HYVs are generally planted in dense stands with high plant populations, which tend to greatly increase the incidence of sheath blight. Relative to sheath blight Crill (1981) stated: "Sheath blight was a minor disease of rice when IR8 was released in 1966. Today it probably causes more loss than any other fungus disease of rice, especially in the lowland tropics." The old landraces that were most important in Asia before the introduction of the HYVs were tall and tended to lodge if heavily fertilized. Before the introduction of the high-yielding rice varieties, populations of the old landraces were low, and thus sheath blight was only a minor problem.

Summary

The density of crop or plant stands has important effects on disease incidence and severity. Dense plant stands generally increase disease, but in some cases (especially with virus diseases) may reduce disease. Crop density can be altered by manipulations of the rate of sowing and planting, in addition to practices such as pruning, thinning, trellising, fertilization, water management, staking, and harvesting plants or plant parts (Palti 1981). Avoiding foliage or root contact can also reduce the incidence of some diseases.

Traditional farmers utilize all of the above practices, but documentation of their use of such practices specifically to manage plant diseases is difficult to find. As with most of the agronomic practices of traditional farmers, plant disease management is seldom a conscious objective. Careful consideration of the value of such practices by traditional farmers relative to the effect they may have on plant disease incidence could shed light on hitherto inexplicable or poorly understood farmer practices.

5

Adjusting Depth of Planting

There is little information found in print on the practices of traditional farmers relative to the planting of seeds or other planting material at the proper depth. However, the uniform and perfect stands one finds of many crops grown by traditional farmers in Asia and other regions is testimony to their knowledge and skill in the proper utilization of this cultural practice.

Shallow Versus Deep Planting

Depth of planting exerts an important influence on the germination and development of plants, especially those from seed. Palti (1981) writes: "One of the periods in which most crops are particularly susceptible to disease is the stage of germination and emergence of seedlings from the soil, up to the time the young stem has hardened to some extent." Deep planting generally lengthens the seedling stage. Examples of the value of shallow planting versus deep planting for disease management are found in the modern plant pathology literature . An increase in the depth of planting increases the amount of several smuts of the seedling types (Neergard 1977). Seedling diseases caused by *Fusarium* spp. and *Rhizoctonia* spp. are also more serious when seed is planted deeply.

The fungus *Rhizoctonia solani* attacks the sprouts of potatoes, and if potato seed pieces are planted too deep, the fungus can completely girdle and kill the the emerging shoots. Planted at a shallower depth, sprouts will emerge; and, even if attacked by the fungus, can survive to produce a productive plant (Rich 1983, Tarr 1972, Walker 1950). Therefore, practices that encourage rapid emergence contribute to the management of *Rhizoctonia*. These practices include warming the potato seed tubers, green sprouting, and planting in warm soil that is not wet. Avoidance of seed tubers with *Rhizoctonia* sclerotia (black fungal resting bodies) is also advisable, although if the soil is heavily infested with the fungus this procedure may not be worthwhile. Rotations are also important in reducing *R. solani* inoculum in the soil (Frank and Murphy 1977).

Leach and Garber (1970) reported that shallow planting contributes to the management of *Rhizoctonia* on beans and sugar beets. Gäumann (1950) reported that when rye was planted too deeply, it was attacked by *Fusarium*, but shallow planting encouraged faster germination and shortened the stage at which plants are susceptible. Greaney (1946) found that the severity of root rot of wheat (caused by *Fusarium* spp. and *Cochliobolus sativus*) increased with the depth of planting.

Traditional Practices

That ancient farmers paid attention to depth of planting is illustrated by the following statement written in 1348 by the Moor Ibn Luyun in Almeria, Spain (Equaras Ibañez 1988): "The soil over the seed should be one to three fingers in depth, or less, and they say that sand should be spread on the planted surface, in order to maintain moisture. Put little soil on delicate seeds, so that the soil above them is light." Bassal (1955) recommended a specific depth of planting for almost all of the crops of importance in Islamic Spain in the eleventh century. For example, he suggested that chick peas and faba beans should be planted at a depth of "two fingers".

Jones and Sif-El-Nasr (1940) reported that in Egypt farmers had different strategies for planting cereals. Seed broadcast on moist fields and then plowed under (germinating at an average depth of 8 cm) had a high incidence of smut, but seed sown on dry land and immediately irrigated (average depth 4 cm) had less damage. Seed broadcast one hour after flooding (surface planted) had the least disease. Covered smut of barley (*Ustilago segetum* var. *hordei*), wheat bunt (*Tilletia tritici*), flag smut of wheat (*Urocystis agropyri*), and covered kernal smut of sorghum (*Sporisorium sorghi*) all responded in a similar fashion.

Summary

Few examples of traditional practices altering depth of planting for plant disease management are to be found in print, but this should not lead one to conclude that traditional farmers are not cognizant of the importance of the proper depth of planting for seeds and other propagating materials. The perfect stands found in many traditional systems testify to skill in manipulating depth of planting. In my experience, traditional farmers in Colombia knew the proper depth to plant crops and were surprised to find that everyone did not possess such basic knowledge. Shallow planting is often an effective practice as plants are especially susceptible to disease during germination and emergence of seedlings from the soil. Depth of planting does affect disease and should be considered when designing disease management strategies.

6

Adjusting Time of Planting

Time of planting is of paramount importance to traditional farmers, as it has such a considerable effect on plant yield. In extreme cases, it may mean the difference between abundance and famine. The date that a traditional farmer chooses to plant a crop may be influenced by many factors. For example, past experience, traditions, superstitions, phase of the moon, magic, existing and expected weather, and advice from family and neighbors may all be considered. According to Morley and Brainerd (1946), Mayan priests selected the dates for burning and planting *milpas* (slash and burn fields) by using their astronomical knowledge. The conditions of the flowers of a cactus are considered in the highlands of Bolivia for determination of the best dates for planting potatoes (Hatch 1983). Lewandowski (1989) wrote that maize planting by the Seneca Iroquois began "when the leaf of the oak or dogwood was the size of a squirrel's foot or ear." Buffalo Bird Woman of the Hidatsa Indians in North Dakota stated that her people knew when it was time to plant maize by observing the leaves of the wild gooseberry bushes, the first in the woods to leaf in the spring (Wilson 1987). Numerous examples of planting date selection based on plant development are found in print.

It is important for traditional farmers to plant on dates that ensure a steady supply of food during the year. Whether farmers are producing food for family use or for the market also has a major influence on scheduling decisions. Historically, a major responsibility of the government and priests of many civilizations was to advise or mandate planting dates and other agricultural activities from past experience and their knowledge of astronomy.

Decisions on planting dates in rainfed areas are generally much more critical than decisions in areas with access to irrigation. Temperatures at the time of planting obviously become increasingly important in the selection of planting dates in the higher latitudes.

In the arid and semi-arid tropics, the hot seasons may increase, reduce, or eliminate many pathogen and vector populations. Rains in the hot, humid rain forests may also increase, reduce, or destroy certain pests and pathogens. Farmers take advantage of pathogen population fluctuations by sowing on traditional dates, which they have found by experience to be optimal for stable yields. In recent years pesticides have often eliminated or reduced the importance of these traditional sowing practices by allowing repeated plantings without serious disease problems, at least for a few years.

Effect of Planting Dates on Plant Disease

Palti (1981) wrote: "The choice of sowing dates in relation to plant diseases has one principal aim, viz. to reduce to a minimum the period over which infective agent (propagule, vector) meets susceptible host tissue." Stevens (1960) and Palti (1981) reviewed the choice of planting dates as they affect disease management. Diseases can sometimes be avoided by planting at times of the year unfavorable for disease development. For example, dry seasons are unfavorable for the germination and penetration of the spores of a fungus such as *Colletotrichum lindemuthianum* (causal agent of anthracnose of beans). Thus, by planting beans so that they will develop during the dry season, the disease can be avoided.

Planting dates are thus of considerable importance in plant disease management. In many cases early or late sowing allows plants to escape pathogen attack. Plants are often more susceptible at certain stages of development, and they may be susceptible to a given pathogen in the seedling stage and resistant at a later stage. Examples were given by Dickson (1947). Early planting in temperate regions reduced fusarium scab of barley (*Gibberella zeae*), as infection occurred most commonly when soil temperatures are high. However, *Gibberella fujikuroi* attacked maize in early spring, causing a seedling blight. Thus, late planting can help to reduce maize seedling blight, as seedlings were most susceptible during germination and the early seedling stage, especially if the soil was cold.

Kozaka (1965) cited numerous Japanese references indicating that early planting of rice reduced damage from rice blast (*Pyricularia oryzae*). He suggested that there is less blast on early planted rice in Japan because of low temperatures sub-optimal for fungal infection at tillering time when rice plants are at their most susceptible, and high temperatures at heading, which are optimal for fungal infection, a critical time for neck rot infection.

A'Brook (1964, 1968) wrote that early planting contributed to the management of the groundnut rosette virus and its aphid vector (*Aphis craccivora*). Dense stands of peanuts inhibited the landing response of the vector, as fewer aphids landed on the close-spaced plants that resulted when groundnuts were planted early.

Traditional Manipulations of Planting Dates

The writings of many ancients, e.g. the Romans (Cato 1934, Varro 1934, and Columella 1988) and the Incas (Poma de Ayala 1987), gave detailed instructions on what agricultural labor needed to be done and what to plant for each month of the year. The correct choice for time of planting has been a major concern of farmers throughout history. Bassal (1955) recommended a time of planting for almost all of the crops of importance in Islamic Spain in the eleventh century. Ibn al-Awam (1988), a Spanish Arab writer of the twelfth century, gave the recommended time for planting for hundreds of different crops.

In the tropics there are far more opportunities to adjust planting dates to escape pathogens or pathogen vectors than in temperate regions, where fewer adjustments to planting dates can be made. Irrigation allows considerable flexibility in planting dates, in comparison to primarily rainfed areas. The ability of traditional farmers to stagger planting dates and simultaneously cultivate several crops in intercropping situations is an indication of their ability to schedule plantings. Wilken (1987) described the complicated scheduling of such mixed plantings by Mexican farmers. However, little was found in the literature indicating that traditional farmers schedule plantings in a conscious effort to manage plant diseases.

Wilken (1987) noted that the use of seedbeds, common among traditional farmers in Mexico and Central America, allows optimal scheduling of time of planting. He added that seedbeds require intensive labor, although they may reduce the time that crops are in the field. Scheduling time of planting becomes highly complicated in multiple cropping systems and dooryard gardens. Wilken noted: "In dooryard gardens, complex spatial arrangements are paralleled by equally complex scheduling. Planting takes place continuously, and harvesting of seed, stalks, leaves, roots, tubers, flowers, and fruits is a never-ending process."

In China, planting crops earlier or later than the normal is sometimes practiced for disease management. For example, Williams (1981) wrote that, although Chinese cabbage is normally planted about August 5, by delaying planting until late August or early September, bacterial soft rot and mosaic virus problems were reduced. Stripe rust of wheat (*Puccinia glumarum*) is the most important rust of wheat in China. Late planting of winter wheat was used there to reduce the chance of fall infection (Chiu and Chang 1982).

Navarro (1903) and Antón Ramirez (1865), citing Maria Amor, wrote that farmers in Spain knew that when chick peas were planted early or exposed to cold temperatures, they were generally exposed to serious attacks of *Ascochyta rabiei* (*Ascochyta* blight). Apparently the farmers thought that the problem was due to sunlight acting on dew or rain drops on the leaves, as if the drops were a lens, and causing a burning of the leaves. Farmers went so far as to draw a cord across fields early in the morning to shake dew or rain drops off. Their association of time of planting and dew or rain drops with the disease incidence was correct, but their ideas on the causal agent were not. For centuries, most farmers in the Mediterranean area planted chick peas in the spring to avoid Aschochyta blight, rather than in the winter when yields would be greater in the absence of *Ascochyta rabiei*.

In Pangasinan province, Philippines, early planting is used to avoid infection by maize downy mildew (*Peronosclerospora philippinensis*) (IRRI 1979). Mohamed and Teri (1989) suggested that date of planting is of great importance to traditional farmers as a means of avoiding diseases. They give examples of farmers' practices in Tanzania, reporting that the farmers surveyed stressed that on late planted beans there is a high incidence of bean rust (*Uromyces appendiculatus*), angular leaf spot (*Phaesariopsis griseola*), and heavy aphid infestations.

Summary

Proper selection of planting dates is of great importance in the management of many plant diseases. Moisture or temperature may either inhibit or increase pathogen activity; adjusting planting dates accordingly may allow plantings to escape pathogen attack. Traditional farmers' choice of planting dates is influenced by a variety of factors: biological, environmental, social, and economic. Although scant literature was found on the selection of planting dates by traditional farmers specifically to reduce disease incidence, this important cultural practice should be carefully considered in all disease management schemes.

7

Site Selection

Site or habitat selection is of considerable significance in the management of plant diseases. Because of their different climatic conditions, diseases may be avoided or reduced by choosing to plant in different sites, regions, or altitudes. Within the farm, sites may be chosen or avoided because of previous crops, soil type, air or water drainage conditions, or previous history of disease.

The Greek botanist Theophrastus (372-285 B.C.) wrote in his "Historia Plantarum" that cereals in elevated fields, or those that are wind-swept, are less severely attacked by "rust" than cereals in low-lying fields in valleys (Orlob 1973). Bassal (1955) and Ibn al-Awam (1988), writing in Southern Spain in the Tenth and Twelfth Centuries, respectively, mentioned plants useful as site indicators, showing whether soil was good or bad for crops. Varro (1934) a Roman living in the first century B.C., wrote that site selection within the farm was of utmost importance. Varro wrote the following regarding what to plant and where to plant it:

> For some spots are suited to hay, some to grain, others to vines, others to olive, and so of forage crops, including clover, mixed forage, vetch, alfalfa, snail clover and lupines. It is not good practice to plant every kind of crop on rich soil, nor to plant nothing on poor soil; for it is better to plant in thinner soil crops which do not need much nutriment, such as clover and the legumes.

The climatic variation occurring in the tropics makes site selection of especially crucial importance. Elevations in Colombia, for instance, range from sea level to over 5,700 meters. Some of the wettest sites on earth (e.g. the province of Choco with over 10 meters of rain annually) are only short distances from hot, dry deserts like those found in the province of Guajira. Snow-capped mountains, such as the Nevada de

Santa Marta (5,780 meters), overlook banana plantations at sea level where annual mean temperatures reach 29° C.

One of the most important factors in mountain agriculture is altitude, which has profound effects on the climate. According to Wellman (1962), mean temperatures are approximately 1°F (0.5.6°C) cooler for every 325 ft. (99 m) of elevation, and this is equivalent to traveling about 100 miles (160 km) towards the north in a temperate country. In Peru ancient Indian farmers planted crops in the Andes from sea level to high elevations over 4,200 meters (Murra 1960, Poma de Ayala 1987). Rainfall is highly variable in the Andes. The Pacific coast of Peru has almost no measurable precipitation, while the Pacific coast of Colombia to the north records over 10 meters annually. This tremendous climatic variation obviously has significant effects on the nature of agricultural systems and plant diseases. Traditional farmers often manipulate this variation to their advantage.

A slash-mulch system in the hot, wet coast of Colombia near Tumaco was described by Finegan (1981). Because of high rainfall, the farmers slashed vegetation, but could not burn it. Crops planted in the slash/mulch system are maize, cassava, sugar cane, beans, fruit, trees for wood, taro, sweet potatoes, yams, and tannier. These traditional farmers also utilized plants as "site indicators" for determining the degree of soil fertility, drainage conditions, and the amount of shade present in a potential slash/burn field. They also knew plants that indicated when land was ready for replanting. The selection of the most propitious site or habitat for crop production can significantly reduce disease incidence.

Altitude Manipulations in Mountainous Areas

Murra (1960) described what he calls "vertical control" of different ecological zones by peasants in the South American Andes. Numerous authors have studied "verticality" in the Andes (Masuda 1985, Mayer 1985, Murra 1960, 1968). Hatch (1983) noted that it is customary for Bolivian farmers to have plots at high, intermediate, and low elevations to reduce their climatic risks. The practice of having plots of land at different elevations spreads their work load, gives farmers a diversified diet, and reduces the risk that any given crop will be totally destroyed in a specific field. The practice may also reduce the risk of serious losses due to plant diseases or other pests.

Potatoes in the Andes are grown from sea level to an altitude of 4,500 m (Gade 1975). Traditional farmers often know intimately the qualities of the many potato cultivars they grow, and therefore grow cultivars at different altitudes according to their zone of adaptation. Ewell et al.

(1990) and Mayer (1979) described this altitudinal distribution of potato varieties in Peru. Farmers may also have a number of separate plots in different locations and at different elevations with differing environmental conditions. As extreme examples, a farming community in Bolivia was described by Carter and Mamani (1982) in which the average traditional family owned 21 different plots in the potato-growing areas. Most families in a Bolivian study managed at least 20 plots, while managing 30 was not uncommon (Hatch 1983). Brush (1981) cited Carter as stating that some families in the Irpa Chico area of Bolivia had 90 or more crops spaced in four or five different climatic zones. Traditional farmers in the Andes of Peru also often had dispersed fields, as Rhoades (1988) noted: "I have known farmers who have as many as 90 tiny fields scattered over a valley and frequently located several days' walk apart." These scattered fields would certainly provide some protection against plant disease epidemics.

Asexually propagated crops such as cassava, potatoes, yams, sugar cane, and sweet potatoes in the lower elevations of the tropics frequently become heavily infected with pathogens, especially virus diseases. Traditional growers in Colombia obtained their potato seed from certain high altitude areas where aphids were scarce (such as the town of Une near Bogotá), and therefore virus transmission minimal. Writing about Peru, Rhoades et al. (1988) stated: "In the low zones, the percentage of purchased seed is higher and the producers change and renew their seed more often than in the higher zones, owing to what farmers call *cansancio* (tiredness) or *degeneración* (degeneration) of the seed." Farmers in the Cañete Valley on the coast of Peru obtained their potato seed from growers at high elevations such as the Mantaro Valley and Huasahuasi (Ewell et al. 1990).

The levels of resistance of different potato cultivars to diseases such as late blight of potatoes (*Phytophthora infestans*) are known by some traditional farmers in the Andes, and I suggest that this may enter into their decisions about the altitude at which to plant a specific variety. Late blight hardly occurs at the highest elevations where potatoes are planted, because of the extremely cold temperatures, thus susceptible cultivars could be planted there without serious losses from *P. infestans*.

Another Inca practice that reduced losses due to disease was the placement of storages at high altitudes in the Andes, where low temperatures would prevent deterioration of stored potatoes and grain. Padre Cobo (Mateos 1956) stated: "The placing of these storages at high altitudes, was done by these Indians in order that the contents of the storages were protected from water, humidity, and rotting." Extensive Inca storages were found near Cuzco and Huancayco in Peru, cities at high elevations with very cool temperatures.

Storey (1936) reported that cassava almost free of African cassava mosaic disease was grown in the highlands of Kenya, and that farmers in the lowlands often got their seed from these high altitude areas. The disease is transmitted by whiteflies, which are scarce in cool, high altitude areas.

Stripe rust (*Puccinia glumarum*) is a major disease of wheat worldwide, especially in cooler areas. It is especially serious in the high elevations of Mexico and the Andes of South America (Orjuela 1956, Rajaram and Campos 1974, Rupert 1951). From five years of observations Orjuela (1956) concluded that in Colombia wheat grown at 3,000 m (10° C average temperature) was more seriously damaged by *P. glumarum* than wheat grown at 2,000 m (18° C). Stem rust of wheat (*Puccinia graminis tritici*) is also a problem in Colombia; but, although damaging at intermediate altitudes, the fungus does not cause serious losses at the higher elevations where temperatures are cooler. I suggest that some traditional farmers in the Andes take advantage of this differential susceptibility and choose their cultivars accordingly.

Summary

The selection of a site or a habitat for planting is one of the more important decisions that any farmer makes, and is often significant in the management of plant disease. Different regions or altitudes can sometimes be selected for a crop; within the farm, sites may be chosen to avoid disease because of previous crops, soil type, air or water drainage conditions, or previous history of disease. In some traditional societies, priests and other leaders had a major responsibility to mandate or advise on planting sites, especially when land was shared by the larger community.

Although in print there is little by way of specific evidence, it is hypothesized that some traditional farmers in mountainous areas utilized plots of land at different altitudes to their advantage for the management of plant diseases. In addition, the practice spread their work load, gave them a more diversified diet, and reduced risk from other pests and the vagaries of the weather.

8

Using Clean Seed

The use of healthy seed and propagating material is basic to sound agriculture practice, and is often considered a form of sanitation. Palti (1981) writes: "The two aims of sanitation are to prevent the introduction of inoculum into field, farm, or community, and to reduce or eliminate inoculum from diseased fields." Both true seed and asexual propagating material can introduce pathogens into previously uninfested fields or exacerbate existing disease situations. The list of seed-borne diseases is a long one (Agarwal and Sinclair 1987, Dykstra 1961, Hollings 1965, Mink 1981, Neergard 1977, Noble and Richardson 1968, Palti 1981). All of the major plant pathogens -- fungi, bacteria, viruses and virus-like pathogens, and nematodes -- are sometimes transmitted by seed or propagating material, but fungi and viruses are the most commonly transmitted. Completely disease-free seed is commonly sought, but almost never achieved.

Selection of the best, healthy seed for planting or food was a concern of ancient traditional farmers. Columela (1988) cites Virgil (70-19 B.C.) as suggesting that cereals would degenerate if the largest seed was not selected, one by one, for planting each year. The Spanish Moor Ibn al-Awam (1988), in his book on agriculture written in the twelfth century, devoted an entire section to careful seed selection and to the criteria one should use when selecting wheat seed. Seed should be of good weight, brightness, hardness, plumpness, color, and should not be soft inside. Also, Ibn al-Awam wrote that if seed had an unpleasant smell it should be assumed to be corrupted. The bad smell might have been due to storage fungi or even a disease such as stinking smut of wheat (*Tilletia foetida*), which has an unpleasant odor.

Poma de Ayala (1987), a Peruvian Indian writing in the sixteenth century, described maize seed selection during the time of the Inca empire. After harvest, the highest quality seed was saved for planting, a somewhat lower quality was saved for consumption, and still lower quality seed was used for making *chica* (a fermented drink). Other maize seed was further described as being empty (*maíz vacio*) or wormy (*maíz agusanado*). Mt. Pleasant (1989) describes the careful maize seed selection practices of the Indians of the Iroquois nation. Even pre-Inca civilizations in Peru apparently had large, state-organized storage networks in addition to the stores of individual farmers (D'Altroy and Harstorf 1984, D'Altroy and Earle 1985, De la Vega 1966).

In Wilson's (1987) book "*Buffalo Bird Woman's Garden*" the selection of maize seed was described by Buffalo Bird Woman as follows: "When I selected seed corn, I chose only good, full, plump ears; and I looked carefully to see if the kernels on any of the ears had black hearts. When that part of a kernel of corn which joins the cob is black or dark colored, we say it has a black heart. A kernel with a black heart will not grow." Buffalo Bird Woman was born in 1839 into the Hidatsa tribe in North Dakota and was interviewed by Wilson when she was about 70 years old.

Traditional farmers often have their own sources of "clean" seed. Although their seed might not pass inspection using modern methods of disease detection, for practical purposes traditional practices often produce seed or propagating materials that will give good yields. Only in recent decades have a few traditional farmers had access to certified healthy seed or propagating materials.

Seed Beds

In the twelfth century the Spanish Moor Ibn al-Awam (1988) suggested that wheat and barley be germinated in a seed bed before planting. The seed beds should be prepared with soil of high quality, improved with thoroughly rotted manure, and frequently watered. Once the seedlings appeared, the number of healthy and diseased seedlings should be noted, and only healthy seedling planted in the field. Ibn al-Awam wrote: "And after the seed has germinated, one counts the number of plants germinated to determine the number of healthy seeds as distinguished from those corrupted." The determination of percent germination and the subsequent selection of healthy seedlings for planting or transplanting just described is clearly a disease management measure. Similar procedures were suggested by Ibn al-Awam for seeds of flax, hemp, onion, radish, turnip, cabbage, and other vegetables.

Traditional farmers in Central America usually prepared seed beds with great care (Wilken 1987). Small raised beds were often used for seed beds, and moisture was carefully controlled. Mulches were frequently used to shield the seedlings and soil from the sun and rain, reduce temperatures, and conserve moisture. Farmers in the Choco of Colombia used soil in old canoes, raised off the ground on stilts, for rice seed beds in order to prevent ant damage (West 1957). Much of the rice grown in Asia is first grown in small, carefully tended seed beds before transplanting, and healthy seedlings are selected for transplanting (Figure 8.1). Worldwide, traditional farmers start many vegetables in small seed beds. Wilken (1987) points out that this practice produces a considerable saving in field space. Seedlings are most easily cared for and can be observed frequently in small seed beds. Optimal moisture, fertilizer, and light can be provided. A variety of tropical shrubs and trees such as coffee, cacao, rubber, tea, and citrus are also started in seed beds. An initial screening for freedom from disease and other pests at the time of selection for transplanting is an important traditional practice for disease management.

The use of muck to prepare seed beds in the *chinampas* of Mexico was described by Wilken (1987) and Coe (1964). A layer of semiliquid muck from surrounding canals was spread over the *chinampa* surface. The muck was then cut with knives into small rectangular blocks called *chapínes*. Seeds or other propagating materials were carefully planted in holes made with fingers or sticks in each *chapín*. The *chapínes* were subsequently transplanted to the soil of the *chinampas*, thus giving the crops a good start. Mulches often protected the seedlings and soil from heavy rains. Each *chapín* was carefully examined before transplanting, and only healthy plants were transplanted. Thus, the use of *chapínes* was a method of reducing the problems with damping off and seedling diseases. Wilken (1987) wrote that the beds were 2-3 meters wide with low earthen borders containing the muck. The semiliquid muck was poured to a depth of 3-5 cm for most vegetables and 8-10 cm for maize. Wilken noted that "farmers rate muck deposits on color, possibly on odor, and to a large extent on texture. The silky fine *bien molido* (well ground) muck was neither lumpy or grainy, and was avidly sought for seed beds." Muck was scooped up from the canals surrounding the *chinampas* with long-handled scoops, and transported to *chinampas* in boats.

Traditional Seed Growing Areas

In the tropics, sexually propagated crops such as cassava, potatoes, yams, sugar cane, and sweet potatoes frequently become heavily infected with pathogens, especially viruses and systemic pathogens. This also occurred in Europe and the United States when potatoes were introduced in the sixteenth century. We now know that this "running out" of potato varieties was due primarily to infection with viruses, which caused varieties to yield less as they were used in successive years (Large 1962). Large also noted that potato growers in England had to get "clean" seed from Scotland long before viruses were known in order to avoid "potato degeneration". A similar degeneration took place in Peru as noted by Rhoades (1988): "In the low zones, the percentage of purchased seed is higher and the producers change and renew their seed more often than in the higher zones, owing to what farmers call *cansancio* (tiredness) or *degeneración* (degeneration) of the seed."

Recharte (1989) wrote that traditional Quecha-speaking farmers of the Cuyo-cuyo district of Peru obtained their potato seed from higher altitudes (the town of Untuca). With modern, improved cultivars now available in Peru, the process has been reversed in the Mantaro Valley of Peru. Mayer (1979) wrote:

> Since every farmer knows that sooner or later the potato seed "gets tired" and must be replaced, a system of seed exchange has evolved. Improved or hybrid varieties diffuse gradually from the lower into the higher zones in a process of exchange and adaptation. The altitudinal limits of the improved varieties are thus reached through a process of trial and error.

The Peruvian potato growers of the Mantaro Valley now obtain clean seed from government sources and plant breeding programs. Monares (1979) and Ewell et al.(1990) stated that farmers in the Mantaro valley sold seed of modern cultivars to growers on the coast. Farmers in the Cañete Valley on the coast of Peru obtained their potato seed from producers at high elevations in Peru such as the Mantaro Valley and Huasahuasi. Thus, the exchange of potato seed goes in both directions. Traditional growers in Colombia in the 1950s obtained their potato seed from certain high altitude areas (such as the town of Une near Bogotá) where aphids were scarce and, therefore, virus transmission was minimal.

Storey (1936) reported that cassava almost free of the cassava African mosaic virus was grown in the highlands of Kenya, and that farmers in the lowlands often got their propagating material from these high-altitude areas. The disease is transmitted by whiteflies, which are scarce in cool, high-altitude areas.

Seed of various legumes and cucurbits that are affected by seed-borne diseases, characteristically transmitted in wet weather, is often produced in arid areas under furrow irrigation. For example, bean seed is commonly infected by many different seed-borne pathogens. In the US most bean seed is produced in dry, furrow-irrigated areas of the west, where there are relatively few bean diseases, in order to avoid infection by seed-transmitted pathogens (Guthrie et al. 1975). Halo blight of beans (*Pseudomonas syringae* pv. *phaseolicola*) has been essentially eliminated by producing bean seed in arid areas of Western US (Grogan and Kimble 1967). The International Center for Tropical Agriculture (CIAT) in Central America has shown that healthy seed can produce exceptional yield increases. They state (CIAT 1975): "Clean seed initially provided by CIAT raised yields among 80 small farmers in the Las Monjas and San Matias valleys of Guatemala from 515 to 1,545 kg/ha in a single season." This result illustrates what can be accomplished by providing clean seed to traditional farmers.

Traditional Seed Treatments

Ancient peoples used a variety of different materials to treat seed before planting. Some were of dubious value, but perhaps others had some useful effects. Orlob (1973) lists wine, ashes, urine, ox gall, and amurca (which contains olive oil) as products used by ancient Romans for treating seed. Oil has been used extensively in modern agriculture for plant disease control (Calpouzos 1966, Martin and Salmon 1931). See also Chapter 2. Seed treatment is a common practice in modern agriculture and utilizes various methods, including chemicals, heat, and anaerobic cold water (Newhall 1955, Sharvelle 1979, Stevens 1960).

Cutting Tubers in Colombia

Almost all traditional potato growers in the Andes of South America plant whole seed (tubers) rather than cut seed, which is commonly used in the US (Figure 8.2). Cutting seed is well known to be an excellent way to spread pathogens (especially bacteria and viruses), but it is possible to use cut seed in the US because of seed certification programs and strict sanitation practices. Nevertheless, problems due to the use of cut seed still cause serious losses in the US. As described in Chapter 1, when attempts were made to cut potato seed in Colombia, losses from soilborne pathogens and from bacterial blight (*Pseudomonas solanacearum*) were severe (Thurston 1963). Our program finally returned to using whole seed; a practice traditional farmers knew was practical for their conditions. Colombian farmers probably had discovered over the centuries that cut seed would not produce a crop. We scientists had to rediscover what the traditional peasant farmers of Colombia already knew.

Summary

Clean seed or healthy propagating material, or such material treated to kill pathogens, often has positive and dramatic effects on plant health and crop yield. Clean seed may prevent the first introduction of inoculum into an area, and if pathogens are already present, can reduce disease losses. Fungi and bacteria are the most important seed-born pathogens, but all types of pathogens can be seed transmitted.

Traditional farmers have used several practices that help to manage seed-born pathogens. The extensive use of seed beds and the subsequent transplanting of carefully selected healthy seedlings are examples. Traditionally, potato seed in the Andes was obtained from cool, high-altitude areas where aphids, which transmit many plant viruses, are absent or few. Another practice used in the Andes, utilizing whole rather than cut seed, prevents severe losses due to fungi and bacteria, which occur as a result of infection when tubers are cut (Figure 8.2). Traditional farmers in most of the world used no chemical seed treatments until the last few decades.

Spectacular increases in yield can be obtained by providing farmers with clean seed. As the individual farmer can seldom produce quality seed without outside help, the costs and benefits of providing such seed must be carefully considered by each society. Healthy planting materials should always be utilized when feasible.

Land Preparation Practices

9

Fallow and Disease Management

Fallowing has been practiced for thousands of years. In what is now Iraq, the ancient Sumerians, one of the world's oldest civilizations, practiced fallowing for cereal fields (La Placa and Powell 1990). A sabbatical year was mandated in ancient Jewish religious law; in that for one year farmers were not allowed to plant crops. Exodus 23:10-11 reads: "For six years you shall sow your land and gather in its yield; but the seventh year you shall let it rest and lie fallow." The ancient Romans (Garcia-Badell y Abadia 1963, Spurr 1986, White 1970), Chinese (Youtai 1987), Incas (de la Vega 1966, Poma de Ayala 1987), Mayas (Harrison and Turner 1978), Arabs (Bassal 1955), and many other peoples historically used fallowing to a greater or lesser degree as one of their major agricultural practices. Garcia-Badell y Abadia (1963) wrote that most Roman land was planted to crops one year, and the next year the land was fallowed. If the land was exceptionally rich, or fertilizer was abundant, fallow was not always practiced. The word "fallow" may be derived from an Anglo-Saxon term *fealewe*, which apparently indicated the color of bare or unploughed ground (Wrightson 1889). Some North American Indians knew the value of fallowing. On the floodplains of the Missouri River the Hidatsa Indians normally fallowed for two years. As the master gardener Buffalo Bird Woman said: "Everyone in the village knew the value of a two years' fallowing" (Wilson 1987).

Fallowing differs from rotation in that generally crops are not planted in a fallowed field. Palti (1981) notes that there are several types of fallowing found in agriculture: dry fallowing, wet fallowing (intermittent irrigation for short periods), and flood fallowing (Figure 9.1). Flood fallowing is discussed in detail in Chapter 11. Dry fallowing reduces pathogen populations by plowing or disking under pathogen infested residues and exposing them to drought and heat.

Fallowing is a highly effective cultural practice for managing many pathogens, especially soilborne fungi and nematodes, and many examples could be given. For example, clean fallow (plowed fields maintained without crops or weeds) in cotton fields, and an interruption of continuous cotton plantings, has significantly reduced losses due to Verticillium wilt (caused by *Verticillium dahliae*) in the Southern United States. Clean fallows, combined with rotations of 2-3 years, are also highly effective in controlling wilt caused by *V. albo-atrum* (Schnathorst 1981).

Fallow in Slash and Burn Agriculture

Slash and burn systems using fire have been in existence since the Neolithic era (Conklin 1961). At one time slash and burn was a widespread practice in temperate areas, but now it is primarily used in the tropics. Plots are partially cleared from the forest growth, the cut vegetation dries and is burned, and crops are planted in the ashes. Conklin (1957) defines shifting (slash and burn) agriculture as "any agricultural system in which fields are cleared by firing and are cropped discontinuously (implying periods of fallowing which always average longer than periods of cropping)." Plots can be used for several years and then gradually abandoned to natural vegetation for fallow periods of up to 20 or more years. In the 1700s De Torquemada (1969) and Clavigero (1974) described the fallow periods for slash and burn used by Mexican Indians. Posey (1985) noted that in the Amazon fields are not necessarily "abandoned" after two or three years; rather, the old fields often continued to bear produce of some crops for years. The Kayapó Indians that he studied in Brazil returned for sweet potatoes for four to five years, yams and taro for five to six years, cassava for four to six years, and papaya for five or more years. Bananas continued bearing for 15-20 years. Thus, as crops are harvested from the slash and burn plots long after natural vegetation begins to return, fallow is often incomplete in the slash and burn system.

There are numerous variations of the slash and burn system. The slash and burn, swidden, or shifting cultivation system consists more of a rotation of fields than a rotation of crops. Usually, successive crops of the same species are harvested until the plot is left to fallow or abandoned because weeds have become unmanageable. The type of vegetation or mixture of vegetation during the fallow and the length of fallow would have differing effects on disease severity. The slash and burn system is discussed in greater detail in Chapter 10.

Nematode Management by Traditional Fallowing Systems

Fallowing is often more effective in reducing pathogen populations in combination with crop rotations. An example of this follows. The origin of potato-cyst nematodes is generally accepted to be the South American Andes (Brodie and Mai 1989). In Peru, before the arrival of the Spanish, farmers of the Inca empire used fallow and rotations for potatoes, according to Garcilaso de la Vega, El Inca (de la Vega 1966) and Felipe Guaman Poma de Ayala (1987). The use of fallow by the Incas was also mentioned by De Murúa (1987). Rowe (1963) wrote: "Under the Inca, the family lots were redistributed each year to ensure equality of opportunity and a proper rotation of crops." The Inca empire had a highly organized system of land tenure and access to land. The annual redistribution of land to farmers by Inca officials was based on previous use. Poma de Ayala (1987) lists 32 different Inca classifications of agricultural land, including land for fallow and for rotation. Murra (1987) wrote that the Spanish, in order to justify their taking over Indian lands, suggested that the amount of land in fallow and in the annual Inca redistribution process indicated that there was lots of unused land in the Andes and that the concept of land as property did not exist.

Rotations of six to eight years are still used today by traditional communities in the Andes. Brush (1977), Mayer (1979), Mayer and Fonseca (1979) and Rengifo Vasquez (1987) indicate that traditional Andean farmers not only fallow, but use other crops in their rotations. Brush (1977) describes a typical rotation/fallow as follows:

> A third stratagem used by Uchucmarcan peasants to assure a potato harvest is to cultivate fields for only one to three years before returning them to a long fallow of eight or more years. Farmers usually sow potatoes in the first year and other Andean tubers - oca (*Oxalis tuberosa*), mashua (*Tropaeolum tuberosum*), and ullucu (*Ullucus tuberosum*) -- for one or two subsequent years. The long fallow period lowers subsistence risk in two ways: by reducing the amount of erosion and soil loss and by killing disease vectors such as nematodes and fungi, which remain in the soil and depend on the continued potato planting to survive.

The long crop rotations in the Andes are mentioned by many authors (Blanco Galdos 1981, Brush 1977, Brush 1980, Brush et al. 1981, Camino et al. 1981, Gade 1975, Hatch 1983, Mayer and Fonseca 1979, Orlove 1977, Rengifo Vasquez 1987). Johns and Towers (1981) and Johns et al. (1982) report that mashua (*Tropaeolum tuberosum*), another crop used in rotation with potatoes by ancient Peruvian farmers, contains isothiocyanates, which are nematicidal compounds. Johns et al. (1982) cite an informant as reporting that in Cuyo-cuyo, Peru, local farmers plant mashua interspersed among other tuber crops because they consider mashua resistant to pathogens. Thus, mashua may have acted as a trap crop in the Inca's rotations. Spanish chroniclers cited by Johns et al. (1982) reported that the Incas believed mashua had anti-aphrodisiac activity and fed it to their troops while they were on military operations. In some areas of Peru, men still refuse to eat the tuber.

Brodie (1984) indicates that nonhosts play an important role in the management of the potato cyst nematode (*Globodera rostochiensis*), and states that nematode densities in the soil decline 30-50% annually when a nonhost crop is grown. Thus, the strategy of the Peruvian farmers to rotate with nonhosts (described above) of the potato cyst nematode is a sound nematode management practice.

Through centuries of trial and error the Incas and their predecessors must have learned that this seven or eight year rotation/fallow gave the best potato crops. Studies in Rothamsted, England, demonstrated that a seven year fallow reduced potato cyst nematode populations below their economic threshold so that a profitable crop could be grown (Jones 1970, Jones 1972). We now know that the destructive potato cyst nematodes (*Globodera pallida; G. rostochiensis*) are present in extremely high population levels in most potato growing areas of the Peruvian Andes, since in many areas the traditional long rotation/fallow period is not used. To the Spanish, the Inca system for distribution of land and their fallow/rotation practices probably seemed to be senseless. Long fallow periods were abandoned in many areas of Peru, and serious losses due to the potato cyst nematodes have occurred in Peru ever since their abandonment. Some of the highest populations of potato cyst nematodes in the world are found today in the Andes of Peru.

Ogbuji (1979) and Wilson and Caveness (1980) reported from Nigeria that nematode populations are effectively reduced in the "bush fallow" (slash and burn) system. Nematode populations found in modern systems were larger than those found in traditional slash and burn systems.

Summary

Fallow is often beneficial in reducing losses from plant diseases, especially those caused by soilborne pathogens. As fallow periods provide other agronomic benefits, it is often difficult to determine their precise contribution to disease management. The benefits of fallow have been known since ancient times. Traditional farmers continue to utilize both dry and flood fallowing. Fallowing is generally more effective for disease management in combination with crop rotation. The slash and burn agricultural system, still important in the tropics, utilizes both fallow and rotation, and has significant disease management aspects. An Andean fallow/rotation system, originating with ancient Indian practices, effectively manages the potato cyst nematodes of Peru. The advantages and disadvantages of the use of fallow should be looked at in the improvement of traditional agricultural systems. Both dry and flood fallowing should be considered in planning plant disease management strategies.

10

Fire and Slash and Burn

Fire is one of the oldest tools humans have used in managing agriculture; it has probably been used deliberately since agriculture began (Bartlett 1956, Grigg 1974, Hardison 1976, Johnston 1970, Kayll 1974). Between two and five percent of the globe is burned each year, primarily for agricultural purposes (Monastersky 1988). Fire is the major means for clearing land in forested areas; it figures significantly in the destruction of the rain forests. It affects other important components of the world's ecology. Thus, it is important to understand the use of fire both historically and today.

Historical Use of Fire

Orlob (1973) cites Hopf (1957) as suggesting that a portion of the grain harvest was flamed by prehistoric man as a protection against fungi and other pests, thus accounting for great numbers of carbonized seeds found in archaeological remains. Many ancient peoples used wood ashes as fertilizer (Ibn al-Awam 1988, Columela 1988, White 1970). The Roman Virgil (70-19 B.C.) speculated on the value of fire for agriculture (Lewis 1941) as follows:

> Often again it profits to burn the barren fields, firing their light stubble with crackling flame: It is whether the earth conceives a mysterious strength and sustenance thereby, or whether the fire burns out her bad humors and sweats away the unwanted moisture or whether the heat opens more of the ducts and hidden pores by which her juices are conveyed to the fresh vegetation -- or rather hardens and binds her gaping veins against the fine rain and consuming sun's fierce potency and the piercing cold of the north wind.

Slash and Burn Agriculture

The practice of slash and burn agriculture consists of clearing plots from the forest and allowing the cut vegetation to dry, then burning, and finally planting crops in the ashes. The practice is a means of reducing shade and increasing solar radiation so crops can be grown in the forest. Plots are used for several years and then are gradually abandoned to natural vegetation for fallow periods of up to 20 or more years. Farmers in the Neolithic era used slash and burn systems (Conklin 1961). Salik and Lundberg (1990) cited evidence that Amuesha slash and burn agriculture in the Peruvian Amazon may have been sustained for 4,000 years. They added:

> Under very difficult environmental conditions with rainfall over 6000 mm/year, with acid, infertile, aluminum-toxic soils, and with tropical pest and pathogen pressures, its is no small feat to sustain agriculture. There may be lessons to be learned in the practice of Amuesha agriculture and in the cultural and biological evolution of the agricultural systems.

In addition to "slash and burn," the major English terms used for the system are *shifting* agriculture (Christanty 1986, Conklin 1954, 1957, 1961, De Schlippe 1956, Grigg 1974, Meggers 1971, Miracle 1967, Norman 1979, Nye and Greenland 1960, Peters and Neuenschwander 1988, Posey 1985, Spencer 1966) and *swidden* agriculture (Beckerman 1987, Denevan and Padoch 1987, Dove, 1983, Harris 1971 and 1972, Ruthenberg 1980, Turner 1978a). The terms are used interchangeably in this book. Innumerable local terms for slash and burn are also found in the literature.

Slash and burn agriculture is far more important than most people realize. Dove (1983) states: "According to recent estimates, swidden agriculture is practiced by 240 to 300 million people on nearly one-half of the land area in the tropics." Hauck (1974) wrote that slash and burn agriculture is the predominant method on 30% of the exploitable soil of the world and that it supports 250 million people or 8% of the world's population. Other estimates ranged from 200 to 500 million slash and burn cultivators (Myers 1988c).

Although the system has been severely criticized as destructive, Greenland (1975) wrote: "In the majority of lowland areas of the humid tropics it has been a stable system, providing a limited number of people living on sufficient land as a continuing method of food production, requiring little in terms of inputs." In the introduction to their recent, comprehensive book on slash and burn agriculture in the third world, Peters and Neuenschwander (1988) wrote:

> Increased population pressure and exploitation of tropical forests affect the practice of shifting cultivation. Millions have turned to the ancient system out of necessity, and its inherent sustainability is succumbing to these pressures. This study is a tribute to the system, but a condemnation of its misuse.

Slash and burn agriculture is most commonly practiced today in tropical areas, but has extended in the past into subtropical and temperate zones. Ancient farmers in Europe practiced slash and burn agriculture, and the system was still common in Finland, Russia, and Sweden in the 1800s (Grigg 1974). Pre-European Indian groups in Northern North America commonly used the slash and burn system to grow maize, beans, and squash (Barrerio 1989, Curwen and Hatt 1953, Sturtevant 1974, Swanton 1946, Wilson 1987).

The Franciscan Friar Diego de Landa (1985), who spent 30 years in the Yucatan of Mexico, in 1566 wrote a description of Mayan slash and burn agriculture and the use of pointed sticks for planting, as did Fernandez de Oviedo (1986), in 1526. Mayan priests selected the date for burning *milpas* (slash and burn fields) using their astronomical knowledge (Morley and Brainerd 1946). It was once thought that the Maya relied primarily on swidden agriculture for subsistence, and that the "collapse" of their civilization was due to its destructive aspects. However, in recent years the discovery of extensive relic raised beds, terraces, and evidence of irrigation strongly suggest that the pre-Hispanic Maya were not limited to slash and burn agriculture, and many scholars have abandoned the swidden-collapse thesis (Turner 1978a).

In spite of plentiful rainfall and high solar energy, the lowland humid tropics are not highly productive, due in large measure to difficulties in maintaining soil fertility and managing pests. Over centuries however, traditional farmers developed slash and burn agriculture systems as a solution to the soil depletion problems and as a method for managing pests. Unfortunately, the system often requires as much as 15-30 hectares to feed one person. On steep slopes or under great population pressure, where the number of people the land has to support becomes so great that the fallow periods are greatly reduced, the system can be destructive.

Articles by Beckerman (1987), Posey (1985), and Denevan and Treacy (1987), describing indigenous management of the tropical forests of the Amazon, give insights into the wealth of sophisticated indigenous knowledge and skills of the Bari, Kayapó, and Bora Indians, respectively. The Indians have numerous classifications of different ecological zones and different management strategies for each. The primary staple of the

Bora Indians in the Brazilian Amazon is cassava, but they also cultivate maize, rice, sweet potatoes, cowpeas, peanuts, pineapple, plantain, peppers, yams, sugar cane, and cocoyams in their swidden fields. As noted in chapter 9, in the Amazon fields are not necessarily "abandoned" after two or three years; rather, the old fields often continue to bear produce of some crops for years. The Kayapó Indians that Posey studied in Brazil returned to slash and burn fields for sweet potatoes for four to five years, yams and taro for five to six years, cassava for four to six years, and papaya for five or more years. Bananas continued bearing for 15-20 years. Beckerman noted that "time to abandonment" usually means the time when weeding is stopped, and that plots continue to produce for long periods of time. For example, Pejibaye palms planted in swidden plots bear fruit for decades.

There are innumerable variations on the slash and burn system, and the above is only a generalized outline of what is often an extremely complex system (Conklin 1954, 1957, 1961, Denevan and Padoch 1987, Grigg 1974, Jurion and Henry 1969, Miracle 1967, Sanchez 1976). Miracle (1967) noted that many authors have suggested that the system is simple, whereas an extraordinary diversity of often complex systems exists. The many complicated systems used in West Africa were described by De Schlippe (1956), De Shield (1962), Jurion and Henry (1969), Miracle (1967), and Ruthenberg (1980). Miracle (1967) described a large number of different slash and burn systems used in the Congo basin. Slash and burn is still the dominant traditional agricultural system used in tropical Africa. Eckholm (1976) estimated that a third of the African continent is covered by grassland and that much of this area would still be in forests if fires had not been started by hunters, herders, and farmers. Asian systems were described by Christanty (1986), Conklin (1954, 1957), Kunstadter et al. (1978), Spencer (1966), and Zinke et al. (1978). Grigg (1974) described the use of the system in Africa, Asia, and Latin America.

Swidden plots mimic tropical forest ecosystems in at least two ways that influence pest problems. A great diversity of crops is grown, often as many as 40 simultaneously in one swidden plot (Conklin 1954). This provides a degree of protection because pests and pathogens are seldom able to build up to destructive proportions on the few isolated plants of each species. Also, the shade produced by the closed or partially closed canopy consisting of some trees that are left standing, and tall crop species such as bananas and papayas, reduces the severity of the weed and some plant disease problems. Finally, however, weeds and other pest problems make the cleared plot uneconomic (Nye and Greenland 1960, Sanchez 1976). Valverde and Bandy (1982) wrote that in the Amazon basin it took more than five years for regrowth to eliminate most grassy types of weeds. Weeds are perhaps the major pest problems in slash and burn agriculture. For example, farmers in Nigeria may spend 50% of their time in weeding (Moody 1975). Knight (1978) described a system related to slash and burn called "nkule" used in the grasslands of Tanzania. Grass was collected in piles and soil was placed on top of it. The grass under the mound was subsequently burned. Maize and cucurbits were then planted on the mounds.

Thus, burning, as well as rotation, polycropping, wide spacing, diversity, and shading, are practices that reduce losses from disease and other pests in slash and burn agriculture. Furthermore, the clearing of small plots permits easy migration of biological control agents, such as insect parasites and predators, from the surrounding forest.

Effect of Burning on Plant Diseases

There has been little research on the effect of the high temperatures produced during slash and burn agriculture on plant pathogens. Caveness (1972a and 1972b, IITA 1976) is one of the few researchers who has studied this; his work was concentrated on nematodes. He found that burning a 10 cm leaf litter layer destroyed root-knot nematodes (*Meloidogyne incognita*) to a depth of 9 cm and sheath nematodes (*Hemicycliophora* spp.) to a depth of 19 cm. Soil temperatures were measured at 1, 3, 6, and 10 cm depths. Temperatures reached were 101°, 78°, 54°, and 44° C respectively.

Ewel et al. (1981) measured burn temperatures in a Costa Rican wet forest site. Although surface temperatures of 200° C were found, the mean at a depth of 3 cm was only about 38° C. The burn killed 52% of the seeds and 27% of the species present beneath the mulch prior to the fire. Temperatures reached during burning depended on such factors as the quality and amount of the slash layer (fuel) burned and its moisture content. In Thailand Zinke et al. (1978) measured temperatures of 600° C at the surface and 75° C at a depth of 2 cm under slash and burn. Under a more intense reburn, temperatures of 150° C were recorded at a 5 cm depth. Lal (1987) provided additional information regarding temperatures achieved by burning in slash and burn systems in the tropics. Differences among the reported temperatures are probably due to variations in the amount and quality of the material burned, its moisture content, wind velocity, the duration and intensity of the burn, and soil type.

Nye and Greenland (1960) mentioned and Jurion and Henry (1969) illustrated the burning of piles of dry branches several feet high in the chitemene system in Zaire. Allen (1965), Gourou (1966), Norman (1979) and Wilson (1941) also described similar systems of intense burning of large piles of foliage. In the chitemene system of the Bemba in North-eastern Zambia, branches and slashed vegetation were stacked and burned (Strømgaard 1985). The trees from which vegetation was collected were not completely felled, but lopped and chopped, and thus they regenerated rapidly. Crops were planted after burning these huge piles of vegetation. Strømgaard estimated that about 11 metric tons of collected firewood was burned on a 0.1 hectare area. The chitemene system is apparently used for the express purpose of creating fertile fields. Soil sterilization under such conditions would probably be quite thorough and deep, and many soil pathogens would likely be eliminated.

In experiments at Pullman, Washington, Cook (1986) found that about 75% of the *Pythium* spp. inoculum was eliminated from the top 5 cm of soil by burning a 30 cm thick layer of wheat straw placed on the soil surface. He found that wheat yield was about 20% greater after burning. Ash alone gave no yield response. Cook did not recommend this practice for Washington farmers because of the labor involved and the loss of organic material, but his results illustrate the potential value of burning for disease management.

Shekhawat et al. (1988) describe a slash and burn practice called "*Jhuming*" used by tribal people in the eastern hills of India. The incidence of bacterial wilt (*Pseudomonas solanacearum*) was found to be negligible in *Jhum* lands. When the investigators experimentally burned straw to simulate slash and burn before planting in potato fields at three different locations, they found a 100% reduction of bacterial blight. No details were given on the intensity of the burn or the quantity of materials used.

Conklin (1957) and Ewel et al. (1981) reported that burning kills weed seeds. A number of references note that microbial populations decreased after slash and burn (Ahlgren 1974, Christanty 1986, Laudelot 1961, Meiklejohn 1955, Miller et al. 1955, Moreno, 1985, Nye and Greenland 1960, Sanchez 1976), but little information is available specifically regarding either short or long term effects of slash and burn on plant pathogen populations.

Hardison (1976) reviewed the use of fire to manage plant diseases, and noted that relatively little had been written on the subject. He wrote that fires in forests can have either beneficial effects (i.e. management of fusiform rust caused by *Cronartium fusiforme*) or adverse effects such as predisposing conifers to root rot caused by *Rhizina undulata*. Controlled burning of over one million acres of forests occurred annually in the Southern United States for the control of unwanted vegetation and several major tree diseases.

Hall et al. (1972) reported that North American Indian tribes burned blueberry bushes to improve the next harvest. According to Hardison (1976) lowbush blueberries are still commonly burned in Northeast United States and Canada, and this practice helps to manage several blueberry diseases. Markim (1943) reported the use of fire to manage *Septoria* spp. on blueberries in Maine. Over 50 Indian tribes in North America practiced intentional burning of grasses to increase the yield of seeds of wild grasses and "weeds" used for food (Bartlett 1956, Hardison 1976). Disease management was probably an important result of the Indians' practice.

Controlled burning of grasses to manage diseases and produce clean seed is widely practiced today in the Pacific Northwest. Hardison stated: "Burning of grass fields, now the most single important cultural practice in grass seed production in the Pacific Northwest region of the United States, was started originally for disease control 28 years ago." The practice of burning various grasses for the production of high quality grass seed in the Willamette Valley of Oregon was described in detail by Hardison (1976). Controlled burning helped to manage diseases such as ergot (*Claviceps purpurea*) and the blind seed diseases of rye grass (*Gloeotina granigena*), and has essentially eliminated the seed nematode (*Anguina agrostis*) in addition to other important diseases, insects, and weeds.

Hardison (1976) wrote that burning of rice stubble and straw was common in Vietnam and was believed to eliminate inoculum of *Pyricularia oryzae* (rice blast), *Gibberella fujikuroi* (foot rot), and *Magnaporthe salvinii* (stem rot). In California, burning of straw was considered the most effective method of reducing inoculum for the management of stem rot of rice (*M. salvinii*). In the Philippines burning of rice stubble was recommended for management of sheath blight (*Thanatephorus sasakii*). Thorough burning of rice crop residues has been recommended for management of the rice nematode *Ditylenchus angustus* (Luc et al. 1990).

The primary purpose of burning, as used in the management of plant diseases, is to eliminate inoculum of various pathogens. In some cases this gives complete control, but in many cases control is only partial. Burning is still used by farmers today for management of many plant pathogens. Hardison's (1976) review article listed over 60 examples of the use of fire to reduce inoculum in forest crops, fruits, ornamentals, cotton, potatoes, small grains, and grasses and forages. For example, Lahman et al. (1981) found that burning potato leaves after harvest reduced inoculum of *Alternaria solani* (early blight) and reduced infection of potato tubers harvested from the field. They concluded: "burning, if correctly done, is a practical and effective method of control."

Summary

The Indians of the Amazon, West African farmers, and the mountain people of the Philippines, have had no personal contacts, yet the similarity of their slash and burn agriculture practices is striking. The use of fire is known to be important in the management of plant pathogens, but the effect of burning on plant pathogens in slash and burn systems is rarely mentioned. For example, no mention of plant diseases was found in the recent book by Peters and Neuenschwander (1988) on slash and burn agriculture in the developing world. It is not clear whether the authors ignored the subject or simply assumed that pests were eliminated by burning.

From the examples they considered, Cook et al. (1978) suggested that the amount of pathogen inoculum destroyed by burning was not sufficient to offset the loss of organic matter caused by burning. Their examples were taken from modern, not traditional agriculture; nevertheless this observation should be kept in mind when considering the use of fire for plant disease management.

The positive effects of burning have been the major emphasis in this discussion. The many well-known negative effects of burning (loss of nutrients, environmental contamination, addition of carbon dioxide to the atmosphere, erosion, destruction of fallow species propagules which slows forest regeneration, predisposition to diseases) should not be overlooked when contemplating the use of burning in agriculture. Considering the importance of slash and burn agriculture and the widespread use of burning as a disease and weed management measure, it is amazing that from the millions spent to help traditional subsistence farmers in developing countries, only a few substantive articles were found examining this important pest management measure.

11

Flooding and Plant Disease Management

Flooding has affected most of the globe at one time or another, as over 70% of earth's land surface is covered by once submerged soils or sediments (Ponnamperuma 1972, Beven and Carling 1989). The extent, impact, and causes of flooding have been reviewed by Kozlowski (1984). Flooding can bring disasters or be of great benefit to humans. For millennia, flooding periodically enriched the soils in many areas of the world, such as the deltas of the Tigris and Euphrates, the Nile River in Egypt, the Amazon, and the immense river deltas in Asia. Major civilizations evolved in such river delta areas. The ancient Sumerians, in what is now Iraq, flooded their fields for a considerable time before plowing and also puddled them with oxen (Maekawa 1990). Deep-water rice varieties, which can elongate as floodwaters rise, evolved in floodplain areas and are still used today in Bangladesh, India, Thailand, and Vietnam. Many muck and peat soils used today for agriculture are periodically flooded. Floodplains often combine fertile soils, abundant aquatic fauna, and ease of transportation and communication in tropical areas (Hiraoka 1989). By rendering soils anaerobic, flooding eliminates or reduces populations of soilborne pathogens, thus contributing to higher crop yields.

The indigenous peoples of the Northern Great Plains of the US have been typically represented as warlike, nomadic Indians mounted on horses. However, sedentary farming Indians also lived for centuries on the floodplains of the Missouri River in what is now North and South Dakota. The Mandan, Hidatsa, and Arikara Indians once constituted powerful tribes that controlled almost the entire Missouri River Valley. Their agriculture lasted over seven centuries and was described orally by Buffalo Bird Woman, a Hidatsa Indian, in a book written in 1917 by

Gilbert L. Wilson (1987). Hidatsa agriculture took advantage of periodic flooding by the Missouri river that renewed the soil annually, adding nutrients to it. In addition the flooding probably reduced populations of soilborne organisms that may have attacked their major crops: maize, beans, squash, sunflowers, and tobacco. Utilization of floodplains for agriculture by Indian tribes in the Southern United States was also common (Hurt 1987).

Today, worldwide, over 220 million hectares of land are irrigated, although not always flooded (FAO 1986). Fifty percent of the arable land in China (47 million hectares) is irrigated (Han 1987a), and much of this irrigated land is periodically flooded (Grist 1975, De Datta 1981).

The Paddy Rice System

The paddy system of growing rice is one of the oldest uses of flooding for plant disease management, as traditional farmers have flooded their rice paddies for millennia (Figure 11.1). It is perhaps our oldest sustainable agricultural system. Rice grown in this system has annually produced a ton or two of paddy per hectare, without high inputs of fertilizer or pesticides. Chinese writings in 3,000 B.C. mentioned rice as an important food (Grist 1975), but it may have been first domesticated in Thailand about 3,500 B.C. (Grigg 1974). Ponnamperuma (1972) describes paddy soils for rice cultivation as follows:

> Paddy soils are soils that are managed in a special way for the wet cultivation of rice. The management practices include: a) leveling of the land and construction of levees to impound water ; b) puddling (plowing and harrowing the water-saturated soil); c) maintenance of 5-10 cm of standing water during the 4-5 months the crop is on the land; d) draining and drying the fields at harvest; and e) reflooding after an interval which varies from a few weeks to as long as 8 months.

Although weed control is probably the major benefit of the rice-paddy system, some authors have noted the value of flooding rice fields for the management of plant pathogens (Cook and Baker 1983, Glass and Thurston 1978, Kelman and Cook 1977, Stolzy and Sojka 1984). Flooding reduces the number of fungal propagules, insects, and nematodes in the soil and, by controlling weeds, which may harbor rice pathogens and insects, reduces disease and insect damage.

Rice blast (caused by *Pyricularia oryzae*) is less severe on flooded paddy rice than on upland rice. One of the reason why disease is less severe is that fewer hours of dew occur in paddy than in upland rice, and thus *P. oryzae,* which requires free moisture for spore penetration, is less serious due to the shorter infection periods under flooded conditions. Also, plants are more susceptible when grown under "dry" conditions and more resistant under flooded conditions (Ou 1972, Kahn and Libby 1958). Recently, Bonman et al. (1988) found that a water deficit during the vegetative stage of upland rice increased the severity of both leaf and neck blast of rice. Similar results were obtained in greenhouse studies by Gill and Bonman (1988). Thus flooding may work in several ways to reduce the severity of rice blast, one of the most serious diseases of rice worldwide.

Kelman and Cook (1977) suggested: "The practice of flooding fields for paddy rice and the use of organic material as fertilizers are apparently key factors in the general absence of soilborne diseases in China." Cook (1986) stated: " The use of paddy rice in a rotation is among the most effective natural treatments ever discovered for control of pathogens in the soil." Cook (1981) described the method used in China to manage fusarium wilt of cotton (*Fusarium oxysporum* f. sp. *vasinfectum*). Paddy rice was grown every three or four years in rotation with cotton, and this flooding reduced fungal inoculum in the soil. Credit for the development of these practices clearly belongs to ancient traditional farmers.

King (1926) described how soils in rice fields were converted to raised beds on which a variety of vegetables and other crops were grown. More recently Williams (1979, 1981), writing on agriculture in China, reported that rice was grown for two to three years and then vegetables were grown on raised beds prepared from soil in the rice paddies. The flooding appears to reduce or eliminate populations of soil pathogens, so that vegetables can be grown on the raised beds without serious root disease problems. An almost identical system using raised beds after growing rice was described in Taiwan (Su 1979). Systems strikingly similar are used in West and Central Africa. According to the International Center for Tropical Agriculture (IITA 1988), there are 85 million ha of inland valleys in sub-Saharan Africa, and 80 % of the inland valley fields found in this region practiced an annual cycle of mounding (a type of raised bed) for vegetables, cassava, or sweet potatoes during the dry season and flat tillage for rice in the wet season. In both the Chinese and African systems, in the construction and destruction of raised beds, organic matter and soil nutrients are recycled through incorporation of crop residues and weeds, and, by flooding, populations of many pests and pathogens in the soil are eliminated or reduced, so that crops can be grown without major problems with soilborne pathogens.

Flooding and Disease Management

Flooding has been used for insect (Newhall 1955) and weed management (McWhorter 1972). Stover (1979) noted that fungi, bacteria, and actinomycete populations declined in flooded soils. The anaerobic or near anaerobic conditions produced by flooding are known to reduce populations of many fungal and nematode soil pathogens (Baker and Cook 1974, Guzman et al. 1973, Palti 1981, Stolzy and Sojka 1984). Stover (1979) and Stolzy and Sojka (1984) reviewed the literature on the effects of flooding on plant disease. Flooding can be used to manage plant diseases, but it also can predispose plants to infection by various plant pathogens. In addition, many pathogens can be carried from place to place in flood water (Stover 1979).

Cook and Baker (1983) discussed possible mechanisms involved in biological control relative to flooding. They noted that flooding was not always completely effective, and suggested that some pathogens, such as those that have propagules able to float (for example the sclerotia or resting bodies of *Rhizoctonia solani*, which causes sheath blight of rice), may survive in a rice paddy. Also, algae may produce oxygen in rice fields, which aids in the survival of fungi at the soil-water interface of a flooded field. Algae may also produce nitrogen, which in turn may affect rice diseases (e.g. high nitrogen makes rice more susceptible to *Pyricularia oryzae*). It should be noted however, that algae play an important role in the nutrition of rice in Asia. Alexander (1977) states:

> In vast areas of Asia, rice has been produced for centuries with no known addition in the form of manure or chemical fertilizers. The nitrogen would thus seem to be derived from the air over the paddy field and, as rice itself cannot utilize N_2, the nitrogen gains were assumed to be associated with free-living microorganisms. The results of careful experimentation have revealed that an increase in bound nitrogen occurs frequently in waterlogged soils containing an abundant blue-green algal bloom, and there exists little doubt of the significance of the blue-greens in the nitrogen economy of paddy soils.

The benefits of natural flooding to plant disease management are illustrated by the increased severity of white rot of onions (*Sclerotium cepivorum*) in Egypt since the construction of the Aswan dam in the 1960s. According to Coley-Smith (1987), before 1965 onions were a major Egyptian crop with 160,000 tons exported. In 1978 only 30,000 tons were exported. The decline in production is attributed to soil contamination by *Sclerotium cepivorum*, as prior to the construction of the Aswan dam soils were flooded for three months every year, and this left a layer of uncontaminated soil on land used for onion production. Coley-Smith suggested that the anaerobic conditions produced by the flooding may have reduced the number of viable sclerotia (fungal resting bodies) of *S. cepivorum* in the soil. Presently, 70% of the land in Egypt with potential for producing onions is infested with *S. cepivorum*.

Some bacterial diseases are also managed by flooding. Management of the Moko disease of bananas (*Pseudomonas solanacearum*) by flooding was reported by Stover (1979). Van Schreven (1948) suggested that growing flooded rice for two years eliminated *P. solanacearum* (Granville wilt) and made tobacco culture possible. Thung (1947) also reported that flooded soils had less *P. solanacearum*. Nesmith and Jenkins (1985) reported decreased populations of *P. solanacearum* in flooded soils in their studies. However, in Sri Lanka, Seneviratne (1976) found that flooding rice paddies did not eliminate *P. solanacearum*. Andrews (1937) and Massey (1931) reported that black-arm of cotton (*Xanthomonas campestris* pv. *malvacearum*) was reduced in the Sudan by flooding harvested cotton fields and their infected debris.

Flooding fields has been shown by the following authors to reduce or eliminate populations of various fungal plant pathogens:

Butterfield et al. , 1978	*Verticillium dahliae*
Cook and Baker, 1983	*Fusarium oxysporum* f. sp.*vasinfectum*
Ioannou et al., 1977	*Verticillium dahliae*
Moore, 1949	*Sclerotinia sclerotiorum*
Stoner and Moore, 1953	*Sclerotinia sclerotiorum*
Stover, 1954, 1979	*Fusarium oxysporum* f. sp. *cubense*,
van Schreven, 1948	*Phytophthora nicotianae*

Flood Fallowing for Panama Disease Management

Extensive commercial use of flooding for plant disease management has been described by Stover (1954, 1955, 1959, 1979). In terms of financial loss, fusarial wilt of bananas (caused by *Fusarium oxysporum* f. sp. *cubense)* once ranked among the most important tropical diseases. Fusarial wilt caused huge losses to the banana industry in Latin America in the first half of the twentieth century, when the banana cultivar "Gros Michel" was widely grown. Dunlap, according to Stover (1962), was the first to use flood fallow successfully for the management of banana wilt. The use of the flooding system increased rapidly from 1945 to 1955 in Honduras and Panama. By 1956 about 6,000 ha in Honduras and 10,000 ha in Panama had been flood fallowed and planted. Periods of 3-18 months were tested for the period of flooding, but it was found that 6 months were as good as 12 or 18 months. Later work indicated that if water was drained and the land was plowed 3-4 months after flooding, and then the land was flooded again for 1-2 months, better control (lasting 4-5 months) was obtained. However, recolonization by the fungus was also rapid, and with the rising costs of labor, engineering, and equipment the use of flood fallow became uneconomic. Nevertheless, flood fallow helped to maintain large areas in banana production until resistant "Cavendish" cultivars could be planted. Flood fallow can only be used in areas where abundant water is available and level land is suitable for the construction of dikes, lakes, and water retention. From 0.6-1.5 meters of water was needed for flood fallow to be effective against the propagules of the fungus *Fusarium oxysporum* f. sp. *cubense.*

Stover's 1954 studies showed that the survival of the fungus in flooded soils depended on the temperature of the water, its degree of aeration, and the location of the fungus propagules in the soil (those on the surface of the submerged soil survive longest). He concluded that the amount of oxygen available to the surface oxidized soil layer determines survival of the fungus. Relative to oxygen Ponnamperuma (1984) stated: "Within a few hours of flooding, microorganisms and roots use up the O_2 present in the water or trapped in the soil and render submerged soil practically devoid of the gas." Ponnamperuma (1972) and De Datta (1981) have discussed at length the complex physical, chemical, and electrochemical changes that occur when soils are flooded.

Flooding for Nematode Management

Flooding has often been used for nematode management , and many references are found on its use (Kincaid 1946). Celery fields on organic soils in Florida were flooded to control nematodes (Guzman et al. 1973). Rodriguez-Kabana et al. (1965) reported that hydrogen sulfide produced by anaerobic organisms under flooded conditions killed nematodes. In Surinam and the Ivory Coast nematicides were ineffective where the soils were heavy clays and mucks, but flooding these soils for 45 days to 6 months reduced nematode injury and increased yields of bananas for three years after flooding (Stover and Ostmark (1981). Luc et al. (1990) wrote: "The recommended control of *Meloidogyne* on rice depends on the species. Flooding of soil even for relatively short periods will control *M. incognita, M. javanica* and *M. arenaria* and probably *M. salasi,* but continuous flooding would be necessary for *M. oryzae* and *M. graminicola*."

Land in the Philippines that was naturally flooded once per year had fewer nematodes (Castillo et al. 1978). IRRI studies (1978) were made of fields with 3-year rice cropping histories on different landforms (knolls, river levees, plateaus, plains, side slopes, and bottomlands). They concluded "flooding of lowland rice culture effectively controlled plant parasitic nematodes (*Rotylenchulus* and *Meloidogyne* spp.) in all landforms except river levees." River levees had light, well drained soils and nematode populations were high in these soils even after a crop of rice. Listed below are additional references to the use of flooding for the management of parasitic nematodes:

Brown 1933	*Meloidogyne incognita*
Cralley 1957	*Aphelenchoides besseyi*
Hollis and Johnson 1957	*Tylenchorhynchus martini*
Johnson and Berger 1972	*Meloidogyne spp. and Trichodorus spp.*
Kincaid 1946	*Meloidogyne incognita*
Loos 1961	*Radopholus similis*
Miller 1953	*Meloidogyne incognita*
Rhoades 1964	*Meloidogyne incognita, M. javanica*
Stover and Ostmark 1981	*Radopholus similis*
Sturhan 1977a and 1977b	*R. similis, Ditylencus angustus*
Thames and Stoner 1953	*Meloidogyne incognita*
Wehunt and Holdeman 1959	*Radopholus similis*

The information on flooding has not always been translated into economic management measures. Stolzy and Sojka (1984) go so far as to state " "Flooding soil for long periods has been used in attempting to control nematodes in field soils. This method, however, has usually been unsuccessful." Nevertheless, in the developing countries of the tropics, in some circumstances, flooding may be the most practical and economic practice for managing nematodes, and the practice deserves additional testing and research effort.

Flooding as a Predisposing Factor to Plant Disease

Water is essential for the germination and infection processes of many fungal pathogens (Kozlowski 1978, Yarwood 1978). Many investigators have found that flooding and associated oxygen stress predisposed or increased the susceptibility of plants to infection by various species of fungi such as *Phytophthora, Pythium,* and *Aphanomyces* (Barta and Schmitthenner 1986, Cook and Baker 1983, Kuan and Erwin 1980, Matheron and Mircetich 1985, Mueller and Fick 1987, Stolzy et al. 1965,

Wicks and Lee 1985, Wilcox and Miretich 1985). Miller and Burke (1975) reported that flooding predisposed beans to fusarium root rot (*Fusarium oxysporum* f. sp. *phaseoli*). Ayres and Boddy (1986), Cook and Baker (1983), Kozlowski (1978), Schoeneweiss (1975, 1986), and Stolzy and Sojka (1984) have reviewed the literature on plant water stress as a predisposing factor to plant disease. The possibility of predisposition should be kept in mind when recommending flooding as a plant disease management practice.

Summary

Although traditional agriculture has utilized flooding for millennia for the management of plant pathogens, modern agriculture has not used flooding extensively for managing bacteria, fungi, or nematodes, probably because of the high cost of controlled flooding. The paddy rice system evolved in Asia. In addition to its various agronomic benefits, the system has an important role in reducing soilborne diseases. Although flooding can be used for the suppression of plant diseases, flooding can also predispose plants to infection by various plant pathogens. A major use of flooding in modern agriculture has been for the management of fusarial wilt of bananas, but this practice is no longer used. The continued use of flooding by traditional farmers should be studied, understood, improved upon if possible, and, where appropriate and economically feasible, used as an alternative to pesticides, especially for the management of soilborne organisms.

12

Mulching

Until I began writing this book, I thought I knew what mulching was. In reviewing the literature, I found that mulching means different things to different people. It has been simply defined as "application of a *covering* layer of material to the soil surface" (Rowe-Dutton 1957) or "any *covering* placed over the soil surface to modify soil physical properties, create favorable environments for root development and nutrient uptake, and reduce soil erosion and degradation" (Wilson and Akapa 1983). Webster's dictionary (1960) shows a temperate zone bias and defines mulch as "leaves, straw, or other loose material spread on the ground around plants to prevent evaporation of water from soil, freezing of roots, etc." *Covering* seems to be a key word in most definitions.

Wilken (1987) and Gindrat (1979) distinguished between crop residues, which are developed in situ, and mulches, which include fresh and dried plant material and composts brought to the field (Figure 12.1). However, it should be noted that crop residues are frequently used as mulches. Pathogens are often killed by the heat generated in the production of composts (Hoitink and Fahy 1986). Palti (1981) distinguished between organic amendments (incorporated into the soil) and mulches (what is spread or left, i.e. stubble, on the soil surface).

Unfortunately, mulches provide a good environment for the multiplication and survival of slugs, which sometimes cause serious losses to crops such as beans when mulched. In Costa Rica the same slugs that attack beans also vector a serious human nematode pathogen (Beaver et al. 1984). Mulches may also provide nutrition and a suitable environment for certain plant pathogens. The effect of mulches incorporated into the soil on the C/N ratio is important, as soluble soil nitrogen may be locked up in the microorganisms decomposing the organic material. This may cause a serious nitrogen deficiency, and make some crops more susceptible to soilborne pathogens.

Materials Used for Mulches

The list of materials used as mulches by traditional farmers is very long (Rowe-Dutton 1957, Wilken 1987, Wilson and Akapa 1983). Cereal straw and stalks are perhaps the most commonly used mulches, but other examples are crop debris, sawdust, leaves, grass, corn stover, manure, weeds, reeds, Spanish moss, and various aquatic plants. In modern or commercial agriculture, the list is even longer and includes manufactured products such as various plastic materials, aluminum foil, asphalt paper, glass wool, and paper.

Some authors refer to "live mulches," which are similar to "green manures" (Akobundo 1984, Karunairajan 1982). Live mulches are intercropped with the crop of interest for their mulch value, whereas green manures are also crops grown for their mulch value, but plowed under before planting the crop of interest (personal communication -- M. B. Callaway). The use of green manures is discussed further in Chapter 13 on organic amendments.

Benefits of Mulches

Wilson and Akapa (1983) wrote: "Mulches also decrease soil moisture evaporation, increase infiltration rate, smother weeds, lower soil temperature, and enrich soils." Mulches are especially valuable for protecting seedlings from the impact of rain, hail, and the wind. Mulches can be especially important in tropical areas with heavy rainfall, as they improve water absorption and are important in water conservation. Mulches reduce rain splashing, an important means of dissemination for numerous bacterial and fungal pathogens. Soil temperatures are lower under mulches in warm tropical areas. Valverde and Bandy (1982) cited figures indicating that mulches reduced temperatures in the upper 10 cm of soil by 2° C during hot days and by 5° in the afternoons. Such temperature changes can have have significant effects on the ability of soilborne plant pathogens to cause disease.

Wrigley (1988) cited a number of benefits from mulching coffee with non-living crop residues. He suggested that mulches reduced soil temperatures, protected against rain, conserved rainfall, increased soil nutrients, increased soil organic matter, produced conditions ideal for root growth, reduced weeds, reduced soil acidity, and increased coffee yields. The main disadvantage Wrigley cited for the use of mulches was high labor costs.

Wilson and Akapa (1983) stated that in the tropics crop response to mulching is almost always positive. In Nigeria, Okigbo and Lal (1982) tested 22 different mulch treatments and found that a rice hull mulch increased maize yields by 0.7 t/ha and cassava yields by 12 t/ha. They observed: "As mulches minimize soil erosion, crop yield can be sustained without a requiring bush fallow rotation." They also pointed out the value of leguminous mulches in providing nitrogen to the soil. At the International Center for Tropical Agriculture (IITA) in Nigeria, research indicated that mulch tillage helped to control erosion and weeds and also improved the soil organic matter content (Rockwood and Lal 1974). In Peru, maize benefitted from mulching, but soybeans, peanuts, and cowpeas showed no yield advantage when mulched (Bandy and Sanchez 1986).

Lal (1975) reviewed work on mulching practices in the tropics. Sanchez (1976) also discussed the general use of mulches in the tropics, and Nair (1984) discussed their use in agroforestry systems. One of the major problems with the use of mulches is that large quantities of material are often needed and, unless crop residues produced in situ are used, material has to be brought in from sources outside of the field. This was done in China for centuries, but at a tremendous cost in human labor (King 1926, McCalla and Plucknett 1981, Witter and Lopez-Real 1987).

Effects of Mulches on Plant Diseases

A variety of effects on diseases, positive and negative, result from the use of mulches. Many authors suggest that mulches may reduce the incidence and severity of plant diseases. Rowe-Dutton (1957) made a comprehensive review of the literature on disease management using mulches for a variety of vegetable crops. Much of the information she cited is anecdotal rather than experimental. Mulches contribute to disease management in various ways. Reduction or prevention of soil splashing is an important function of mulches in the management of some plant pathogens (Fitt and McCartney 1986, Gilbert 1956, Galindo et al. 1983b, Moreno and Mora 1984, Rowe-Dutton 1957). According to Fitt and McCartney (1986): "Rain splash is the second most important natural agent, after wind, in the dispersal of spores of plant pathogenic fungi." Intercropping cassava with maize, melons, or other crops reduced soil splashing by rain and significantly decreased the severity of cassava bacterial blight (*Xanthomonas campestris* pv. *manihotis*) in Nigeria (Ene 1977). Mulches served the same purpose as intercropping. Muimba-Kankolongo et al. (1989) found that mulches reduced the incidence of a cassava stem tip dieback of unknown etiology in Zaire. Mulches may prevent direct contact of the foliage, fruit, or vines with the soil and thus prevent diseases transmitted from the soil. Moisture is preserved during dry periods by mulches, so they help provide a constant water supply to plants. The severity of the disease blossom end rot of tomatoes can be reduced by mulching (Rowe-Dutton 1957).

Mulches do not always reduce disease. Experiments in Costa Rica by Mora and Moreno (1984) showed that both incidence and severity of *Stenocarpella* leaf spot of maize (caused by *Stenocarpella macrospora*) was increased by soil treatments that included mulching, compared to those including the removal of crop residues. Bandy and Sanchez (1986), working in the Peruvian Amazon, found that mulches of the grass *Panicum maximum* were detrimental to rice yields, as the rice plants remained green for a longer period and thus were more susceptible to fungal attacks (no fungus was specified).

Cook et al. (1978) have reviewed the literature on the effect of crop residues on plant disease. They list three ways that crop residues left in the field can affect plant diseases:

1. For many plant pathogens, residues provide food and a place to live and reproduce.
2. Residues affect the physical environment occupied by the host and pathogens.
3. As organic soil amendments, residues intensify the microbial activity of the soil and this, along with a variety of decomposition products (some phytotoxic or fungitoxic), may affect pathogens, susceptibility of the host plants, or both.

Organic mulches might have similar effects. Some information is found regarding the use of mulches as a covering layer by traditional farmers, but much more information is found on the use of various organic amendments incorporated into the soil. When organic mulches are so incorporated, there is considerable evidence that they usually have a suppressive effect on plant pathogens and nematodes (Cook and Baker 1983). Organic amendments have been found to be useful in managing nematodes (Castillo 1985, Muller and Gooch 1982, Sayre 1971). According to Hoitink and Fahy (1986) composts prepared with tree barks and used as a mulch may release inhibitors of plant pathogens such as *Phytophthora* spp. and some nematodes. Some organic decomposition products are phytotoxic (Linderman 1970, Patrick et al. 1964), and thus amendments are not always beneficial. As Huber and Watson (1970) observed: "the physical, chemical and biological interactions in soil are so complex and varied that it is a challenge to determine the specific effect responsible for disease control."

Cook and Baker (1983) suggested that organic amendments generally produced enhanced competition among soil microorganisms for nitrogen, carbon, or both, and that this may result in fewer soil pathogen problems. Baker and Cook (1974), Cook and Baker (1983), Garrett (1960), Gindrat (1979), Lewis and Papavizas (1975), Muller and Gooch (1982), Papavizas (1973), and Patrick et al. (1964) have reviewed the effects of organic amendments on soilborne pathogens, but it is clear that additional research is still needed to clarify the overall value of such amendments for managing plant diseases and their future role in agriculture. Gindrat (1979) listed many examples of soilborne pathogens managed by the addition of organic matter, but noted that in many cases huge amounts of organic matter were needed to provide control. There are many soilborne plant pathogens, and the management of each may depend on different site-specific environmental and soil factors.

Traditional Mulching Practices

Traditional farmers, especially the Chinese (Youtai 1987), have used mulches in their agriculture for millennia. It is often difficult to distinguish between mulches, crop residues, organic fertilizer, and green manures, because in much of the literature authors often do not specify whether these organic amendments were used as a covering layer on the soil or incorporated into the soil. Additional examples, besides those that follow, of traditional mulching practices can be found in the chapters on biological control, organic amendments, raised beds, and terraces. Brass (1941) describes the use of raised beds and mulching in the highlands of New Guinea by traditional farmers as follows:

> They (*raised beds*) are made, instead, to get at the rich black swamp deposits and virgin alluvial material of subsurface levels, which, when spread over the impoverished topsoil, brings a new lease on life to the land. But the procedure, as observed, is first to cover the ground with a mattress of cut grass, then to heap the excavated materials on this in a bed 12-15 inches thick.

The Aztec Indians of Mexico used mud from canals, aquatic plants, and manure to spread on the surface of raised beds called *chinampas*. These practices maintained the canals between the *chinampas* and enriched the *chinampa* soils (Coe 1964). In Tlaxcala, Mexico, water hyacinth is collected from ditches and canals and used as a mulch. In addition to its value as a mulch, the process of collecting aquatic weeds and muck cleans the ditches and canals (Wilken 1987). Additional examples regarding the incorporations of organic materials into ridges, mounds, and raised beds are given in Chapter 14.

In Costa Rica, *poró* (*Erythrina poeppigiana*) is a commonly used shade tree for coffee. Trees are pruned 1-3 times a year. The pruned branches provide a mulch and return nitrogen to the soil. Recently, Beer (1988) concluded that *poró*, when pruned 2-3 times per year, can return, as a litter layer, the same quantity of nutrients as are applied to coffee plantations in Costa Rica via inorganic fertilizers, even at the highest recommended rates of 270 kg N, 60 kg P and 150 kg K/ha/yr. In addition, trees contributed 5,000-6,000 kg organic matter/ha/yr. The total leaf litter from both coffee and *poró* was between 5,000 and 20,000 kg/ha/year. This amount is within the range of leaf litter fall reported for tropical forests. Although the nutrient contribution by nitrogen fixation is important, Beer (1988) concluded that, especially in fertilized plantations of cacao and coffee, leaf litter productivity is a more important contribution of the leguminous shade trees than nitrogen fixation. Litter also provides organic material to the soil and shades out weeds. In Ethiopia, when nutrient deficiencies are noted in coffee, leaves and branches of *Erythrina burana* are cut and buried around the coffee bushes. Subsequently, farmers claim higher production for several years (Teketay 1990). Wilken (1987) described the use of leaf litter as a mulch in Guatemala, where the nutrient content of the leaf litter is sometimes enriched by placing it in stables beneath animals. The large quantities of organic matter provided by leaves and leaf litter may have important effects on biological control of soilborne pathogens, but no information was found in this regard.

In the South Pacific farmers also use leaves of *Erythrina* spp. as a mulch. Weeraratna (1990) wrote that farmers there also use grass, weeds, banana leaves, and parts of the coconut palm -- fronds, husks, wood chips, and shredded logs -- as mulches for taro. The application of leaves of *Erythrina* spp. at the rate of 30 tons/ha as a mulch to taro increased yields by 65%. In Uganda, the Ganda farmers maintain bananas in the same field for up to 50 years without rotation by the use of careful pruning, weeding, and mulching (Fallers 1960). Karani (1986) also describes the mulching of bananas in Uganda.

A number of authors (Cieza de Leon 1959, de Acosta 1987, Parsons and Psuty 1975, Ravines 1978, and Soldi 1982), have described the sunken field agriculture used on the coast of Peru and Chile, one of the driest deserts in the world. Crops were grown in excavated depressions, called *hoyas*, close enough to the water table so that plants had adequate moisture. A significant percentage of the land in some coastal valleys of Peru was in sunken fields. Fish were often planted with a grain of maize to provide moisture and fertilizer at the time of planting in these sunken fields (Mateos 1956, Del Busto 1978 and Cieza de León 1985). The use of mulches in sunken fields in Peru was described by Cobo (Mateos 1956), a priest who wrote in the seventeenth century. Indians collected the rotten leaves of a tree called *guarango* and then covered the soil of the sunken fields with a thick layer of the leaves. The purpose of the thick mulch, according to Cobo, was to prevent or remedy the accumulation of salts harmful to agriculture. The thick mulch also probably provided a benefit as an organic amendment. Flores Ochoa (1987) described the cultivation of sunken beds called *quochas* in arid areas as high as 3,840 m above sea level in Peru.

Slash/Mulch Systems

Several early Spanish chroniclers in tropical America described the use of a slash/mulch system for maize, beans, and other crops (Patiño 1956, 1965). Patiño (1965) suggested that Indian civilizations living in humid tropical forests invented the practice. In the sixteenth century Pedro Cieza de Leon in his "Cronica General de Peru" (cited by Patiño 1965) described an Indian practice: "on hillsides they cut the vegetation and plant their roots and other food crops into it." Patiño (1965) also cited Miguel Cabello Balboa in 1577 as reporting another native practice on the coast of Ecuador: "they do no more than broadcast maize seed in the hillsides and cut the vegetation over it and collect the harvest." This certainly is an early description of the slash/mulch method. Patiño cited other descriptions of the practice, including that of Francisco José de Caldas in 1801, from the Choco province of Colombia:

> In those places where it rains continuously such as in the Province of Choco, and the entire west coast of the country, they don't burn; but the excessive humidity combined with great heat makes the land there very fertile. They plant in these areas with no other operation except to cut the small bushes and trees, broadcasting at the same time the grain, after which they cut the vegetation covering the maize.

Conklin (1961), Finegan (1981), and West (1957) referred to the system as slash/mulch. The areas described above, such as the Choco of Colombia, have extremely wet climates. For example, Quibdó, Colombia, located in the Province of Choco, receives over 10 meters of rain annually (West 1957). Patiño cited other descriptions of the slash/mulch system (called *tapado* in Spanish), not only from Colombia, but also from Panama and Costa Rica. After its introduction to the Americas by the Spanish, rice was also planted in the *tapado* system.

West (1957) wrote a fascinating description of the extensive use of the slash/mulch system on the Pacific coast of Colombia (Province of Choco) and in Ecuador:

> Throughout most of the Pacific lowlands, however, the heavy precipitation and lack of a dry season precludes the effective use of fire. Instead a peculiar system, which might be called "slash-mulch" cultivation, of probable Indian origin, has evolved. Seeds are broadcast and rhizomes and cuttings are planted in an uncleared plot; then the bush is cut; decay of cut vegetable matter is rapid, forming a thick mulch through which the sprouts from the seed and cuttings appear within a week or ten days. Weeds are surprisingly few, and the crops grow rapidly, the decaying mulch affording sufficient fertilizer even on infertile hillside soils.

West described the cutting of vegetation in the Choco as a community affair or "*minga*" with ten or fifteen men and women cutting the bush together with their machetes. Maize, cassava, and plantains were crops planted in the Choco using the slash-mulch system. A primitive variety of maize called "*chococito,*" especially adapted to the slash-mulch system, was commonly planted in Ecuador, Colombia, and Panama (Patiño 1956).

A slash-mulch system on the hot, wet coast of Colombia near Tumaco was described by Finegan (1981). In this high rainfall area, farmers slashed the vegetation, since they could not burn it. Maize, cassava, sugar cane, beans, fruit, taro, sweet potatoes, yams, tannier, and trees for wood were planted in the fields. Certain plants were used as "site indicators" for determining the degree of soil fertility, drainage conditions, and the amount of shade present in a potential slash/mulch field.

Carter (1969) described the use of velvet beans (*Stizolobium* spp.) as a mulch by Kekchi Indians in the lowlands of Guatemala. The luxurious growth of the velvet beans may reach a height of 2.5 meters in six months. The Kekchi slashed the growth with machetes and chopped it up finely. A mulch 8-10 cm thick of the decayed velvet bean matter was left on the soil, after which maize was planted. Carter claimed that plots planted to velvet bean did not revert to grassland or forest, and some plots had been used consecutively for 14 years of dry season farming with little indication of diminishing fertility. The above observations, if confirmed, indicate the possibility of sustaining soil fertility in the lowland tropics with the velvet bean or other cover crop systems for long periods of time with a minimum of inputs.

Frijol Tapado System

Traditional farmers in many areas of Costa Rica grow beans (*Phaseolus vulgaris*) using a slash/mulch system called in Spanish "*frijol tapado*", which in English means "covered beans." According to Patiño (1965), Professor Tulio Ospina of Colombia was the first to use the name "*siembra de tapado*" for the practice. The procedure consists of broadcasting bean seeds into carefully selected weeds, then cutting and chopping the weeds with a machete so the broadcasted bean seeds are covered with a mulch of weeds (Araya and Gonzalez 1987, Cavallini 1972, Galindo et al. 1982, Jimenez 1978, Patiño 1965, Skutch 1950, von Platen et al. 1982, von Platen 1985). A semi-determinate type of bean, between a bush and a climbing bean, is planted. The beans grow through the mulch and eventually cover it. This combination of mulch and bean plants effectively prevents weed growth and appears to conserve soil moisture. In addition, the mulch prevents soil splashing, which was found in a Costa Rican study by Galindo et al. (1983a, 1983b) to be the most important source of inoculum of *Thanatephorus cucumeris* causing a severe bean disease called web blight. The fields selected for *tapado* are generally occupied by broadleaf weeds and certain grasses, which will not regrow after they are cut. Thus the weeds do not compete with the beans for light, nutrients, or moisture. The disease is effectively managed by traditional farmers who use the traditional practice of *frijol tapado*, even in areas where climate is optimal for web blight development. Skutch (1950) described the practice as follows:

> The bean seed is broadcast through the low, dense vegetation, which is then cut down with machetes and chopped up (*picado*) so that it lies close to the ground. The bean vines sprout up through the mulch of stems and leaves, finally covering them over. No cultivation of the crop is necessary or feasible.

Web Blight of Beans

Web blight of beans is caused by the fungus *T. cucumeris* (asexual stage -- *Rhizoctonia solani*). The disease was described in detail by Thurston (1984). In the humid lowlands of the tropics, web blight is possibly the single most destructive disease of beans. Beans are traditionally grown in cooler, temperate areas in Latin America, but because of population pressures, farmers migrate from high to low-altitude areas and often take beans with them. In warm and humid tropical areas *T. cucumeris* can cause rapid defoliation of beans and sometimes complete crop failure. In 1980, an epidemic of web blight occurred in the Guanacaste region in the northern part of Costa Rica, resulting in a 90% reduction in bean yields (Galindo 1982). This loss occurred on beans planted under clean cultivation. As with many tropical diseases, precise information on yield losses is difficult to obtain, but the disease has been characterized as severe in Mexico (Crispin and Gallegos 1963), Costa Rica (Echandi 1965), and elsewhere in Latin America (Cardenas-A. 1989, Schwartz and Galvez 1980).

The main sources of inocula that can initiate infection are mycelial fragments and sclerotia (fungal resting bodies). Basidiospores (airborne sexual spores produced by Basidiomycete fungi) can also cause infection (Cardenas-A. 1989, Echandi 1965, Galindo et al. 1983). The study by Galindo et al. (1983) found sclerotia and mycelia free in soil or in the form of colonized debris to be the main source of inoculum in the hot, humid areas of Costa Rica. Inoculation of beans occurred mainly by splashing of rain drops containing infested soil. Large numbers of small sclerotia were produced on rain-splashed soil and debris adhering to bean tissues and on detached tissues on the soil surface. These sclerotia provide new sources of inoculum, which again can be splashed onto beans. Weber (1939) suggested that sclerotia may also be disseminated by wind. In the study by Galindo (1982) in Esparza, Costa Rica, infections caused by basidiospores were observed as previously reported by Echandi (1965). However, the lesions observed were not numerous, remained restricted in size, and apparently caused little damage.

Web Blight Management by the *Tapado* System

The *frijol tapado* system was compared in Costa Rica with another mulch system (a 2.5-cm thick layer of rice husks, a cheap by-product commonly found in the area) using both web blight susceptible and tolerant bean cultivars (Galindo et al. 1983b). Bean yields were increased significantly by mulching with rice husks or by *frijol tapado* (Table 12.1). Rice husks and *frijol tapado* were equally effective in avoiding splashing of infested soil and in managing web blight (Figures 12.2 and 12.3), and both treatments gave better control of web blight than the fungicide pentachloronitrobenzene (PCNB). PCNB is a highly effective chemical against *R. solani* and can be applied as a soil or foliar treatment.

In the absence of web blight, the yields in fields under the *frijol tapado* system are generally lower than those in fields planted in drilled rows with clean cultivation. For this reason, some in Central America oppose continuation of the *frijol tapado* system; however on small farms

TABLE 12.1. Effect of Mulch Treatments on Bean Yield of Two Cultivars Planted in Two Web Blight Infested Fields in Costa Rica in 1980

	Bean seed yield (kg/ha)			
	Experimental field		*Commercial field*	
Mulch Treatment	*Porillo 7*	*Mexico 27*	*Porillo 70*	*Mexico 27*
None (clean cultivation)	0	0	273	217
Frijol tapado	-	-	637	534
Rice husks	655	587	835	679

Source: Galindo, J. J., G. S. Abawi, H. D. Thurston and G. Galvez. 1982. "Tapado", controlling web blight of beans on small farms in Central America. N.Y. Food and Life Science 14(3):21-25. p. 25.

in Costa Rica, most of the beans currently produced are grown using the *frijol tapado* system. Small farmers persist in using the system because of its low risk, its small investment in labor (primarily to cut weeds), and because there is always some yield even when prolonged periods of rain produce conditions that allow *T. cucumeris* to destroy beans under the clean cultivation system. Von Platen et al. (1982) noted that covered beans could be planted on steep hillsides without erosion problems. Also, once planted, *tapado* fields required little if any maintainence, so farmers could safely leave a planting while they went to harvest coffee or engage in other off-farm activities. *Tapado* fields required less labor and, although they have a low productivity per land unit, they have a high return on a labor per work-day basis. Furthermore, the *tapado* beans would suffer less from possible prolonged droughts, as compared to the clean cultivation system, and thus the risk of decreasing bean harvests is reduced.

Studies by Galindo et al. (1983a, 1983b) on management of web blight by mulching indicated that in the area of Costa Rica where they conducted their research, and during that time period, basidiospores played a minor role in disease spread. Cardenas-A. (1989) studied web blight of beans in Colombia and found that at higher, cooler elevations basidiospores played an important role in the disease epidemiology.

Mulching was of no value in management of the disease under the conditions of his experiments (in Darien, Colombia, 1,400 meters above sea level). Cardenas reported that the maximum and minimum temperatures there were 23.6° and 16.5° C, while Galindo (1982) reported that the maximum and minimum temperatures in his study area (Esparza, Costa Rica) were 30° and 20° C, respectively. Rainfall was also much higher in the experimental site in Costa Rica. These climatic differences probably help to explain the different results obtained. This is a striking example of the need for site-specific studies when using mulching as a plant disease management practice.

Other Traditional Practices for Web Blight Management

Galindo et al. (1982) noted that in Costa Rica *frijol tapado* fields were generally planted in hilly areas. Farmers selected hills that received full sunlight early in the morning, thereby reducing the periods of high humidity, which favor the web blight disease. Over 50 farmers were interviewed in Tabasco, Mexico by Rosado May and Garcia-Espinosa (1986) relative to their strategies for management of web blight of beans. In this area yield losses of up to 95% of bean production due to web blight had been recorded. The *tapado* system was used in association with maize, and farmers also increased the distance of planting for better disease management. Two farmers claimed that they saw no web blight in fields where the "good" weed *Euphorbia heterophylla* (painted leaf) had been prevalent. Sadly, all farmers interviewed indicated that they were expecting a chemical solution to the web blight problem.

The *frijol tapado* system is an excellent example of a traditional system that is easily managed, requires low inputs, is sustainable over time and environmentally sound, and, for the farmers that practice it, provides a secure source of food and income that fits in well with their off-farm activities. A challenge remains to modify the system to make it more productive without losing its advantages.

Summary

The use of mulches provides many agronomic benefits, such as lowering soil temperatures, protecting against erosion, providing nutrients and organic matter to the soil, improving soil texture, and reducing weeds. In addition, mulches may manage plant diseases by reducing soil splashing of primary inoculum, influencing the moisture content and temperature of the soil, and enhancing the soil microbiological activity that suppresses soilborne plant pathogens. Mulches may also have negative effects on crops, as decomposition byproducts may be phytotoxic, mulch residues may harbor pathogens or other pests, or the microenvironment, when changed by mulches, may be more favorable to disease development. Nevertheless, the positive benefits of mulches have been generally found to greatly outweigh these potential negative aspects. Effective use of mulches, relative to the control of plant diseases, varies according to location, crop, and pathogen.

Traditional farmers have used mulches for millennia. One of the major problems with the use of mulches is the large quantities of material that are often required. Unless the crop residues or weeds that are produced in situ can be utilized, material has to be transported from outside the field, often at great expense and at a tremendous cost in human labor. Some societies have such labor available, many others do not. In the hot humid tropics where plant growth is rapid and luxurious, there should be many opportunities to increase the use of green manures and natural vegetation as mulches. Considering the potential value of mulches for erosion control, nutrient enhancement, temperature control, weed suppression, and disease management, it appears that there should be far more research on this important subject, using some of the resources expended on agricultural development in tropical developing countries.

13

Organic Soil Amendments

Many traditional farmers, such as those in China (Chandler 1981, King 1926, McCalla and Plucknett 1981), India (Raychaudhuri 1964), Mexico (Coe 1964), Papua New Guinea (Brass 1941, Waddell 1972), Peru (Poma de Ayala 1987), ancient Rome (Columela 1988, Spurr 1986, White 1970), Spain (Bassal 1955, Al-Awam 1988, Bolens 1972), and South and Central America (King 1926, Siemens 1980, Turner and Harrison 1981) have added considerable quantities of organic material to the soil. Youtai (1987) wrote that the value of manures was established in China before the fifth century B.C. According to Spurr (1986), at some times in ancient Rome the quantities of manure used in agriculture were higher than those used in Italy today. One of the tasks farmers had during the time of the Incas in Peru was to carry "much" manure (*mucho estiercol*) to their maize and potato fields in July (Poma de Ayala 1987). Denevan (1987) noted that the soils of Peruvian terraces were "now maintained by the application of manure and compost and by periodic fallowing, and both were likely true prehistorically." Von Hagen (1959) quoted Cieza de León as describing Inca practices in the 1500s as follows: "They bring back the droppings (*guano*) of birds to fertilize their cornfields and gardens, and this greatly enriches the ground and increases its yield, even if it once was barren. If they fail to use this manure, they gather little corn."

Llama trains from the Peruvian coast transported *guano* from birds for fertilizer to the highlands during the time of the Incas (Gade 1975 and Julien 1985), and the Chimu, another ancient Peruvian civilization, used *guano* from islands off the coast in their agriculture (Ravines 1980). Peruvian Indians not only fertilized with *guano*, manure from corralled llamas and other cameloids, but also with anchovies, green manure, and ashes, according to Del Busto (1978). Fish such as anchovies or sardines were often planted with a grain of maize to provide moisture and fertilizer at the time of planting (Cieza de León 1985, Del Busto 1978, Mateos 1956). North American Indians in New England also used fish as a source of fertilizer (Barrerio 1989). Human waste was highly regarded in Peru at the time of the Incas as a fertilizer for maize, and was dried, pulverized, and stored for planting maize (Weatherwax 1954). Cobo (Mateos 1956), a Spaniard writing in the seventeenth century, suggested that the Spanish could learn how to fertilize from the Peruvian Indians.

Orlob (1973) noted that many ancient Indian, Islamic, Roman, and medieval Spanish writers repeated over and over a common remedy for diseases of trees and shrubs; namely to remove soil around the plant and replace it with manure, other organic matter, and/or ashes. Raychaudhuri (1964) gives numerous examples of such treatments used in ancient India. Reading the books written centuries ago by Columela (1988), Alonso de Herrera (1988), and Ibn al-Awam (1988), I also found that application of manure was their most common recommendation for disease management. These authors characterized many diseases as "red leaf" or "yellowing." Many of their "diseases" were probably nutrient deficiencies that may have responded well to these recommendations. Ibn al-Awam (1988), an Arab living in Southern Spain, also wrote that various types of manure were useful in curing similar problems of many crops, such as bananas, apples, peaches, citrus trees, figs, grapes, palms, cedars, and wheat. He gave recommendations for restoring fertility to soil by using various combinations of crop residues, straw, manure, and ashes. Manure was sometimes added to raised beds. Ibn al-Awam also described methods of treating and composting animal manure and human waste. Bassal (1955), an Arab writer of eleventh century Spain, extolled the virtues for agriculture of "much manure". Numerous additional ancient Greek, Roman, Arab, and medieval authors also praised the virtues of manure for agriculture. Composting of manure and use of manure pits was recommended by several ancient authors (Al-Awam 1988, Cato 1934, Columella 1988, Ibn Luyun (Equaras Ibañez 1988), Varro 1934). Thus, the ancients knew the value of organic matter not only as an essential agronomic practice, but also for the management of plant diseases.

Types of Organic Matter

Organic matter is 1) either brought to fields from elsewhere or 2) consists of crop residues or green manure crops, that are incorporated into the soil. Palti (1981) distinguished between organic amendments (incorporated into the soil) and mulches (which are spread or left, e.g. stubble, on the soil surface). Organic amendments used by traditional farmers consisted of manure, composts, aquatic plants, mud from rivers, streams, and canals, and crop and plant debris. Watson (1983) lists the many types of organic matter used in ancient Arab agriculture. Manure from many different domestic and wild animals, in addition to human waste, was used. Animal products such as blood, urine, and powdered bones, horns, and ivory were incorporated into the soil in addition to vegetable matter such as straw, husks, amurca, leaves, rags, shavings, and other plant debris.

Organic matter applied to fields in Asia was described by King (1926) as follows:

> For centuries, however, the canals, streams and the sea have been made to contribute toward the fertilization of cultivated fields, and these contributions in the aggregate have been large. In China, in Korea, and in Japan all but the inaccessible portions of their vast extent of mountain and hill lands have long been taxed to their full capacity for fuel, timber, and herbage, for green manure and compost material; and the ash of practically all the fuel and all the timber used in the home finds its way ultimately to the fields as fertilizer. In some cases organic material had to be transported long distances, and this was time-consuming and expensive.

King (1926) further observed:

> In China enormous quantities of canal mud are applied to the fields, sometimes at the rate of even 70 and more tons per acre. So, too, where there are no canals, both soil and subsoil are carried in to the village and there they are, at the expense of great labor, composted with organic refuse, then dried and pulverized, and finally carried back to the fields to be used as home-made fertilizers. Manure of all kinds, human and animal is religiously saved and applied to the fields in a manner which secures an efficiency far above our own practices. Statistics obtained through the Bureau of Agriculture, Japan, place the amount of human waste in that country in 1908 at 23,950,295 tons, or 1.75 tons per acre of her cultivated land.

Long-term soil fertility is often enhanced by the use of organic sources of nitrogen, although much of the nitrogen supplied by organic sources is not immediately available. Bouldin et al. (1984) state: "The organic fraction of manure has many of the properties of the ideal N fertilizer -- it is not subject to leaching or denitrification losses, it is not toxic to plants, and it mineralizes N at a rate dependent on the same climatic conditions as plant growth." Huge quantities of organic manure are produced in the US. King (1990) estimated that 160 million metric tons were produced in 1979. Although King stated that 90% is returned to the land, much of it is not properly utilized. Describing the situation relative to the use of animal manure in the in the US, Bouldin et al. (1984) wrote: "At least 50% of the manure N is not recycled through the farming system, and there is reasonable evidence that no more that 25% of the manure N from most feedlot, dairy, and poultry operations is recycled."

Organic Amendments and Plant Disease Management

There is an extensive literature on the positive effects of organic amendments on plant pathogens (Baker and Cook 1974, Cook and Baker 1983, Lewis and Papavizas 1975). Most introduced biological control agents against soilborne plant pathogens are added to the soil with organic matter. A classic example of the positive effects of adding copious quantities of organic matter to the soil is given by Baker and Cook (1974) and Shea and Broadbent (1983). In Australia, *Phytophthora cinnamomi* causes a severe root rot of avocados. Growers who add large quantities of chicken manure to the avocado soils have little problem with *P. cinnamomi*. Growers in the same area with little organic matter in their soils have severe problems with avocado root rot. Borst (1986) and Coffey (1984) reported that mulches reduced damage caused by *Phytophthora cinnamomi* to avocado.

The addition of large quantities of organic amendments does not always control soil pathogens. In a few cases amendments may increase disease, at least in the short term (Cook 1986, Garrett 1960, Kaiser and Horner 1980), as some soilborne pathogens may survive on organic amendments. Also, some organic decomposition products are phytotoxic (Linderman 1970), so amendments are not always necessarily beneficial.

As Huber and Watson (1970) point out, "the physical, chemical and biological interactions in soil are so complex and varied that it is a challenge to determine the specific effect responsible for disease control." Cook and Baker (1983) suggest that organic amendments generally produce enhanced competition among soil microorganisms for nitrogen, carbon, or both, and that this may result in fewer soil pathogen problems. Baker and Cook (1974), Cook and Baker (1983), Garrett (1960), Gindrat (1979), Lewis and Papavizas (1975), Muller and Gooch (1982), Papavizas (1973), and Patrick et al. (1964) have reviewed the effect of organic amendments on soil pathogens, but it is clear that considerable additional research is still needed to clarify the overall value of such amendments for managing plant diseases and their future role in agriculture. There are many examples of soilborne pathogens managed by the addition of organic matter, but in many cases huge amounts of organic matter are needed to provide effective management.

Nematode Management

Organic amendments are useful in managing nematodes (Castillo 1985, Rodriguez-Kabana 1986, Rodriguez-Kabana et al. 1987, Rodriguez-Kabana and Morgan-Jones 1988, Sayre 1971). When chitin, available in some areas as crustacean, fish and other animal wastes, is added to soil, there is an increased parasitism of nematode eggs by fungi (Rodriguez-Kabana 1986, Rodriguez-Kabana and Morgan-Jones (1988). Muller and Gooch (1982) list 125 papers referring to the use of various organic amendments for management of nematodes. Rodriguez-Kabana (1986) and Rodriguez-Kabana et al. (1987) have reviewed the effect of adding organic amendments to the soil for nematode management, and this appears to be a promising area for further research, especially in developing countries where the cost of nematicides is prohibitive. Cook and Baker (1983) suggest : "The addition of copious quantities of organic matter has reduced damage from root-knot nematodes in some cases, perhaps because populations of trappers and hyperparasites increase or because the nematodes are attracted to the organic matter rather than to the roots." The addition of cow dung to yam mounds before planting in Ghana increased yields and significantly reduced nematode numbers (*Scutellonema bradys*), according to Adesiyan and Adeniji (1976).

The Use of Organic Amendments in China

Wittwer et al. (1987) estimated that organic sources furnish about half of the nutrients applied to crops in China. McCalla and Plucknett (1981) stated:

> The story of organic fertilizer use in China is a fascinating one. Probably nowhere else in the world have organic fertilizers attained the same level of importance. For centuries, Chinese farmers have labored to gather and utilize human and animal wastes, crop residues, and other organic and inorganic wastes. Many of the practices described here — and taught elsewhere as modern scientific farming -- have been a most efficient system of waste recycling.

Cook and Baker (1983) wrote that about 80% of The People's Republic of China's fertilizer requirements are met with organic sources such as composted crop residues, green manures, human waste, and livestock manure. Over 100 tons of compost is often used on a single hectare of land. They state:

> Perhaps the best large-scale demonstration of effective biological control by cultural practices is the widespread multiple-cropping *organic* system used in the People's Republic of China. The agriculture of that country, which feeds nearly one fourth of the earth's population, clearly demonstrates that farming can be both intensive and sustainable and, if stabilized for years or perhaps for centuries, can provide a biological balance and disease suppression similar in effect to the disease suppression that can occur with prolonged monoculture of some crops.

McCalla and Plucknett (1981) estimated that in 1974 there were 9.5 million metric tons of nitrogen produced in China from organic sources. They described in detail the collecting, transporting, and processing of organic fertilizers. Sources of fertilizers in China included crop residues, green manure, pond, river and canal silt and sediments, soil from uncultivated areas, burned soil, plant ashes, chicken manure, weeds, and aquatic plants. Dazhong and Pimentel (1986) analyzed a seventeenth century farming system in China, and concluded that the system was "generally sustainable" and maintained soil nutrients and organic matter.

The combined practices of flooding fields for rice and using organic matter for fertilizers "are apparently key factors in the general absence of soilborne diseases in China." (Kelman and Cook 1977). Fertilizers that are primarily organic contribute to root health not only by improving soil structure, but also by suppressing or eliminating inoculum of soilborne plant diseases (Kelman and Cook 1977). Cook (1986) suggests: "Improvements in root health, by making the root system more efficient, can contribute as much if not more to growth and yield of a crop as can high rates of fertilizer." In describing Chinese agriculture, Youtai (1987) states:

> Agricultural production methods practiced over thousands of years in China have proved that the application of organic fertilizers or manure is the most effective means to improve soil structure, raise the productivity of the land, and achieve a "sustainable agriculture" even with intensive cultivation. The Chinese-based "organic agriculture" has not only won international recognition in recent years but has a long history of confirmation in practice as well.

Organic Amendments in Mexican Chinampas

Large quantities of organic material were used in the *chinampas* of Mexico (Coe 1964) and are described in more detail in Chapter 14. Mud rich in nutrients from the bottom of the canals was dredged up by hand and spread on the *chinampa* surface. This practice maintained the canals and enriched the *chinampas*. In addition, aquatic weeds, animal manure, and (in the time of the Aztecs) human waste were also spread on the *chinampas*.

As previously described in Chapter 3, Lumsden et al. (1987) studied *chinampa* soils relative to root disease occurrence. When they compared relative levels of damping-off disease incidence caused by *Pythium* spp. on seedlings grown in soils from the *chinampas* with those grown in soils from modern systems of cultivation near Chapingo, Mexico, they found that disease levels were lower in the *chinampa* soils. When they introduced inoculum of *Pythium aphanidermatum* into the *chinampa* soils the fungus was suppressed. From their studies they concluded that the copious quantities of organic matter added to the chinampa soils stimulated suppression of *Pythium* due to biological activity in the soil of organisms antagonistic to the fungus. Zuckerman et al. (1989) also studied suppression in Mexican *chinampa* soils, but of plant parasitic nematodes rather than fungi. They found that the high organic content of the soil is probably responsible in part for the relatively few nematodes in *chinampa* soils, but they also found nine organisms with antinematodal activity.

Organic Amendments in Other Traditional Systems

The Arab Ibn Luyun of Almeria, Spain (Equaras Ibañez 1988), wrote the following in 1348: "The straw of faba beans, barley, and wheat sweetens the earth and improves it greatly, and they say that it is used also against "tizón" of grapes. It is applied in December onto plants that have signs of the tizon, and it eliminates the disease." What disease the "*tizon*" of grapes might have been is unknown, but the organic material appeared to have beneficial effects.

Farmers of Papua New Guinea grew sweet potatoes on especially prepared mounds. Waddell (1972) stated that over 20 kg of sweet potato vines, sugar cane leaves, and other vegetation were placed in the mounds. As the material began to decompose, the mound was closed with soil and subsequently planted with sweet potato cuttings. Waddell noted that diseases did not appear to be a serious problem in these plantings.

Incorporation of organic material into mounds and raised beds in Africa by traditional farmers is common (Miracle 1967). Fresco (1986) and Miracle (1967) mentioned burning organic matter, that had been incorporated into mounds. When cow dung was added to yam mounds before planting in Ghana, yields were increased and nematode numbers (*Scutellonema bradys*) were significantly reduced, according to Adesiyan and Adeniji (1976). Many other examples of the benefits of adding organic matter to the soil by traditional farmers are found in the literature.

Traditional Use of Green Manure Crops

The value of green manure crops in agriculture has been known for centuries. Cato (1934), a Roman who lived 234-149 B.C., wrote that lupines, beans, and vetch fertilized the land. Varro (1934), writing between 116-27 B.C., suggested that some plants, although they gave no benefit the year they were ploughed under, gave benefits the following year. Varro wrote:

> Some crops are to be planted not so much for their immediate return as with a view to the year later, as when cut down and left on the ground they enrich it. Thus is is customary to plough under lupines as they begin to pod -- and sometimes field beans before the pods have formed so far that it is profitable to harvest the beans - - in place of dung, if the soil is rather thin.

Green manure has been used for centuries in China, and its present and past use has been mentioned by Cook and Baker (1983), King (1926), and McCalla and Plucknett (1981).

The practice of using green manure crops has important benefits, as they add organic matter to the soil, and play a role in the suppression of soilborne pathogens while improving the physical condition of the soil. Nutrients may also be added to the soil, especially from leguminous green manures. A wide variety of plants have been used as green manure crops (Karunairajan 1982). *Crotalaria spectablis* is often used as a cover crop and subsequently plowed under as a green manure. Root-knot nematodes (*Meloidogyne* spp.) enter the roots of *Crotalaria,* but do not survive. Thus, *Crotalaria* also acts as a trap crop and can be useful in the management of nematodes.

Palti (1981) noted that a great variety of effects on diseases, positive and negative, result from the use of green manures, and that it is important to take into account the effect of green manure crops on the

C/N ratio as soluble soil nitrogen may be locked up in the microorganisms decomposing the organic material. Some crops may be more susceptible to soilborne pathogens if there is a serious deficiency of nitrogen.

Wilken (1987) wrote that cover crops and green manures are not widely used in Central America and Mexico by traditional farmers. However, he did note one rather unusual system in Ostuncalo, Guatemala. The soils in the area are volcanic, have a high sand content, and are low in nutrients and organic matter. The farmers grew in their fields trees called *sauco* (*Sambucus mexicana*) and pruned them annually, so only stumps were left. Leaves and small branches were chopped up and incorporated as a green manure into the fields, where potatoes, maize, and beans were grown. According to Wilken, farmers of the area claimed that good crops depend on this practice. Carter (1969) described the use of velvet bean (*Stizolobium* spp.) as a combination green manure and mulch by Kekchi Indians in the lowlands of Guatemala. (See Chapter 12).

Contributions Possible from Human Waste

Numerous ancient Arab, Chinese, Greek, Roman, and Spanish authors extolled the benefits of human manure, and some gave specific instructions on how to process it and get a product that was odorless and useful as a fertilizer. One Spanish author (De el Seixo 1793) even claimed that crop yields resulting from the use of human manure were "monstrously" large. Human waste still is widely used in many traditional agriculture systems.

Witter and Lopez-Real (1987) recently calculated that the potential contribution of human waste in the United Kingdom could be nearly 40% of the current demand for nitrogen fertilizer. At present it accounts for less than 3% of the demand. Sewage sludge could also be important, but contamination by industrial pollutants, especially heavy metals, is a major problem. In developed countries, without subsidization, sewage sludge cannot compete economically with chemical fertilizers. They suggested that contamination seems to be a less important problem in developing countries. The lack of effective treatments for human pathogens, and cultural taboos, limit the use of human waste in many developing countries. Unless human waste is properly treated to eliminate pathogens, the many human diseases that such pathogens can cause are a serious concern. However, Witter and Lopez-Real reported that composting is an effective treatment for sewage sludge and night soil, and can produce a hygienic and aesthetically acceptable product.

Hoitink and Fahy (1986) have reviewed the literature on the suppression of plant diseases by composted municipal sludge.

Summary

F. H. King had been a professor of soil science at the University of Wisconsin before he went to China, Korea, and Japan in 1907. His observations on agriculture in those countries make fascinating reading. In today's world, where energy shortages and concern about the environment are increasing, his observations (King 1926) are highly pertinent:

> It could not be other than a matter of the highest industrial, educational and social importance to any nation if it could be furnished with a full and accurate account of all those conditions which have made it possible for such dense populations to be maintained upon the products of Chinese, Korean, and Japanese soils. Many of the steps, phases and practices through which this evolution has passed are irrevocably buried in the past, but such remarkable maintenance efficiency attained centuries ago and projected into the present with little apparent decadence merits the most profound study.

An important conclusion can be made from this chapter. Historically, many sustainable agricultural systems incorporated large quantities of organic matter into soil. Numerous highly respected authorities in the discipline of plant pathology suggest that the incorporation of large quantities of organic matter into soil generally results in reduced soilborne disease in addition to other important agronomic benefits, and the practice should be recommended whenever feasible.

1.1 Spraying potatoes with pesticides in the Bolivian Andes. Misuse of modern pesticides in traditional systems is common today. (Courtesy Robert W. Hoopes)

8.1 Selection of seedlings from a rice seed bed in Taiwan. In Asia most rice is first grown in small, carefully tended seed beds before transplanting, and disease free seedlings are selected for transplanting.

8.2 Planting whole potato tubers in Ecuador. Almost all traditional potato growers in the Andes plant whole seed (tubers) rather than cut seed. Cutting tubers is known to spread pathogens, but this practice prevents their spread. (Courtesy Roger A. Kirkby)

9.1 Fallow land on hillside in Bolivian Andes. Fallowing is a highly effective cultural practice for managing many pathogens, especially soilborne fungi and nematodes. (Courtesy Robert W. Hoopes)

11.1 Puddling a rice paddy with a carabao. The paddy system of growing rice is one of the oldest uses of flooding for plant disease management.

12.1 Jamaican farmer carrying mulch to a scallion field. Fresh and dried plant materials are brought to the field in many traditional farmer systems. (Courtesy Dennis Finney)

12.2 (*above*): Effective control of web blight of beans provided by mulching with rice husks. The practice prevents rain splash that disseminates the propagules of the pathogen. (Courtesy American Phytopathological Society)
12.3 (*right*): Close-up of bean plants mulched with *frijol tapado* debris that prevents rain splash. (Courtesy American Phytopathological Society)

14.1 *Chinampa* field in Xochimilco, Mexico, and boat for collecting muck from canals. *Chinampas* are a specialized raised bed system that includes the incorporation of organic matter into the raised beds. (Courtesy Alan G. Turner)

14.2 Straw mulch covering *chinampa* seed bed, Xochimilco, Mexico. A covering layer of straw is usually placed over the seed bed for protection. (Courtesy Alan G. Turner)

14.3 Irrigating a mulched raised bed near Bangkok, Thailand. Raised beds are commonly used in Asia for agriculture, especially in high rainfall and swampy areas.

18.1 Maize storage in Ghana. Cooking fires are placed beneath the storage to aid in drying. (Courtesy Conrad Bonsi)

18.2 Maize field in Mexico in which ears have been twisted down (*doblando la mazorca*). By using this practice, the grain is protected from rain; it dries better on the plant in the sun than in storage, is less likely to be blown down by the wind, is less accessible to rats and birds, and reaches such a low moisture content that storage deterioration by fungi and insects is greatly reduced.

18.3 Yam storage in Nigeria. These traditional storages allow yams to be kept for over six months.

19.1 Mixed cropping in the Andes of Ecuador. Crops include maize, common beans, faba beans, mustard, and squash. (Courtesy Roger A. Kirkby)

22.1 Trimmed poró (*Erythrina poeppigiana*) trees commonly used as shade in Costa Rican coffee plantings. The trimmed branches provide nutrients and organic matter and produce a mulch that suppresses weeds.

22.2 Dense shade produced under a cassava canopy. Traditional farmers take advantage of this by weeding only until the canopy discourages further weed growth.

23.1 Women weeding rice by hand in Taiwan. Traditional farmers spend an inordinate amount of time in weed control with some spending up to half of their time weeding.

24.1 Seventeen different types of beans grown by a farmer on a one and one-half hectare farm near Puebla, Mexico. The diversity of varieties gave protection against pests and the vagaries of climate. (Courtesy American Phytopathological Society)

24.2 A rain forest in South America. Much of the area in tropical forests is being rapidly lost to cattle pastures.

14

Raised Beds

The management of wetlands for agriculture by raised beds or raised fields has been practiced extensively by indigenous peoples of the Americas and by Chinese farmers for at least 2,000 years. Darch (1983), Denevan (1970), Denevan et al. (1987), and Parsons and Denevan (1967) described more than 170,000 ha of raised field remnants found in South America. Extensive systems of raised fields known as *chinampas* were found in Mexico and were also common in Central America (Adams et al. 1981, Barrera et al. 1977, Gomez-Pompa, 1978, Siemens 1980, Siemens and Puleston 1972, Turner 1974, Turner and Harrison 1981).

North American Indians used raised beds in their agriculture before European arrival in several areas of the US (Fowler 1969, Riley and Freimuth 1979). Raised beds were also common in Africa (De Schlippe 1956, Jurion and Henry 1967, IITA 1988, Miracle 1967) and Asia (FAO 1980, King 1926, Harwood and Plucknett 1981, Hsu 1980, Williams 1981). The practice is quite ancient, as raised bed farming was developed in China in the fifth century B.C. (Wittwer et al. 1987). Raised beds in the Wahgi Valley of New Guinea have been dated as older than 350 B.C. (Lampert 1967). Ibn al-Awam (1988), a twelfth century Arab writer who lived near Seville, Spain, recommended raised beds for green beans, chicory, radishes, onions, melons, lettuce and eggplant. Manure was often added to the raised beds. His recommendations are valid today: "Eggplants placed in soil mounded into raised beds, grow robust and very advanced." Regarding raised beds for lettuce, he wrote: "This manner of planting -- in raised beds -- is very good, as the plants receive water uniformly from below, unlike plants grown on level ground."

In 1778 the Spanish writer Francisco Vidal y Cabasés described a new method for cultivating wheat on raised beds recommended by the Englishman (Jethro) Tull. Vidal y Cabasés wrote that using the raised bed system wheat could be grown for many years without fallow; that the system gave excellent yields while using less fertilizer (manure); and that there was less soil compaction, better distribution of water, and in heavy rains better drainage. His claims of better yields were backed up with data provided by a Señor Thomé of the Royal Society of Leon, Spain. In 1775 and 1756, using rye, Señor Thomé compared the raised bed system with normal practice in two different locations. In one location he obtained 530 pounds of grain from raised beds, versus 174 pounds in the same extension of land planted on the level. In another locations yields were 207 pounds on raised beds versus 73 pounds on the level. In spite of these dramatic results, I found no evidence that the system was used again for grain in Spain, probably because of the heavy labor requirements for making large hills and the difficulty of planting wheat on them. Thus, the benefits of raised beds were known centuries ago, but European civilizations of the 1700s were, I suspect, discouraged by their high labor requirements.

Denevan and Turner (1974) defined a raised field as " an agricultural feature created by transferring earth to raise an area above the natural terrain." Denevan (1970) differentiated the following types of wetland cultivation used by indigenous American peoples:

1. soil platforms built up in permanent water bodies
2. ridged, platformed, or mounded fields on seasonally flooded or waterlogged terrain
3. lazybeds or low, narrow ridges on slopes and flats subject to waterlogging;
4. ditched fields, mainly for subsoil drainage
5. fields on naturally drained land, including sandbars, river banks, and lake margins
6. fields diked or embanked to keep water out.

"Raised beds", "ridged fields", "drained fields", and "cambered beds" (Webster and Wilson 1980) are also terms found in the literature to describe raised fields. Hills, ridges, and mounds, types of broken raised beds, historically have been routinely used for many root and tuber crops. Ibn Luyun (Equaras Ibañez 1988) mentioned *caballónes* (raised beds or mounds) used for trees in the thirteenth century in Almeria, Spain. De Schlippe (1956) described several types of ridges and mounds cultivated by the Zande people of tropical Africa. The different ridging practices of the Kofyar people of the Jos Plateau in Nigeria were described by Netting (1968). For some of their crops the Kofyar used enclosed ridges, called tie ridges, that trap water. After a rain their fields appear to be checkered with pools. Ochse et al. (1961) described the construction of ridge terraces, erosion dikes, and bench terraces, which are used in "modern" tropical agriculture.

Ridge tillage is becoming popular in the US. It is considered to provide significant erosion control benefits. The practice also helps to overcome some of the soil temperature, soil compaction, and weed problems found in many modern systems of agriculture (National Research Council 1989a). Recently, many home gardeners in the US have become interested in raised beds, and books describing their use are available (Carr 1978, Chan 1985, Jeavons 1982, Mittleider 1986).

Relatively little is reported on the utility of raised fields for plant disease management, but there is little doubt that, in addition to their obvious irrigation, drainage, and agronomic value, disease management is often an additional benefit.

Chinampas

Probably the best known raised field system is the *chinampas* or "floating gardens" of the Valley of Mexico, which the Spanish conquistadors erroneously thought floated (Figure 14.1). When the Spanish arrived in Mexico in 1521 and entered the capitol of the Aztec civilization located on an island in Lake Texcoco, they were amazed by the immense areas in *chinampas*. *Chinampas* apparently interfered with the Spanish conquest. As Squier (1858) noted: "The lands around the lake were highly cultivated, as appears from the references made by Villagutierre to 'great fields of maize', surrounded by fences and deep ditches, which were impossible for the Spanish horsemen to leap." Early descriptions of *chinampas* were made in the 1500s by De Acosta (1987) and in the 1700s by Javier Clavigero (1974) and Torquemada (1969).

Despite Spanish attempts to drain Lake Texcoco for flood control, which diminished greatly the area in *chinampas* (Mateos 1956), some are still farmed near Mexico City at Xochimilco (Armillas 1971, Gomez-Pompa, 1978, Jimenez-Osornia and del Amo 1988). Only 1,000 ha of *chinampas* remain, one-tenth of the area the Aztecs possibly had under cultivation 2,000 years ago (Salas 1988). The productivity of the *chinampas* was cited as a major factor that allowed the Aztecs to grow from a small tribe to a powerful group that dominated most of Mexico.

Chinampas were probably first developed by the Maya and then later utilized by other Indian cultures in Mexico and Central America (Adams et al. 1981, Chen 1987, Siemens 1980, Siemens and Puleston 1972, Redclift 1987, Turner and Harrison 1981). Adams et al. (1981) stated "New data suggest that Late Classic period Maya civilization was firmly grounded in large-scale intensive cultivation of swampy zones." Many investigators do not believe that slash and burn agriculture could have provided sufficient food for the large Maya populations, which constructed Tikal, Palenque, and other Mayan centers. Only a more intensive food production system, such as raised field agriculture, could have produced food sufficient for the large populations that existed then. Over 165 square miles of raised fields have been documented in Maya areas (Rice 1991). Pollen data indicated the possibility of raised fields on the Hondo River (between Mexico and Belize) going back to 1800 B.C. Radiocarbon dating of worked timbers in the area dated them at 1110 B.C. plus or minus 230 B.C. (Puleston 1978). A system similar to *chinampas* was being used by Indians on the swampy shores of Lake Maracaibo, Venezuela when the Spanish arrived (Simon 1892).

The *chinampas* currently found at Xochimilco, constructed in shallow Lake Texcoco, are generally rectangular in shape (90 m by 4.6-9.0 m) and separated by canals (Coe 1964). The surface of the *chinampas* is usually a meter or so above the water level in the canals. Two operations build up the *chinampas*. First, mud rich in nutrients from the bottom of the canals is dredged up using a hand tool and spread on the *chinampa* surface. This maintains the canals and enriches the *chinampas*. In addition, aquatic weeds and animal manure (and in the time of the Aztecs, human waste) are spread on top. A wide variety of crops were grown by the Aztecs on the *chinampas* and many diverse crops are still seen today (Muñoz 1986). Maize is planted directly in the *chinampas*, but other crops are first planted in seedbeds prepared by spreading a layer of mud over vegetation, cutting it into small rectangular blocks called *chapines*, and planting a seed in each *chapín*. A covering layer of straw is usually placed over the seed bed for protection (Figure 14.2). The *chapines* are subsequently transplanted to the soil of the *chinampas*, thus giving the crops a good start. The *chinampas* are perpetually moist, and cropping can continue year round, even through the dry season. Although little information on yields was found, according to Armillas (1971) yields of the system are very high. In 1978 Venegas (cited by Redclift 1987) calculated maize yields of 4-6 tons/ha, while Gomez-Pompa (cited by Chen 1987) reported yields of 6-7 tons/ha on *chinampas* in Tabasco .

Chapin (1988) describes the unsuccessful attempts to reintroduce *chinampa* technology into the lowland tropics of Mexico. He is highly critical of these attempts and correctly identifies many of the technical, social, and political reasons for failure. Nevertheless, his strong implication that *chinampas* are not a valid model for increasing food production in tropical areas seems superficial and premature. *Chinampa* construction and maintainence require considerable labor, and this meant that the discovery of immense oil reserves in Tabasco, which increased greatly the cost of scarce labor, poor planning and coordination of government projects, and lack of marketing opportunities and planning contributed more to failure than a faulty conceptual model.

The *chinampa* system made continuous cropping possible by sophisticated water control, multiple cropping, high levels of organic material and nutrients periodically added to the system, and transplanting of healthy, selected seedlings (*chapines*) with strong root systems (Gómez-Pompa 1978, Jiménez-Osornio and del Amo R. 1988). These practices contribute towards good disease management. The diversity of crops grown on traditional *chinampas* may have also contributed to the success of the system by inhibiting disease spread.

Lumsden et al. (1987) studied *chinampa* soils relative to disease. They compared relative levels of damping-off disease caused by *Pythium* spp. on seedlings grown in soils from the *chinampas* versus those grown in soils from modern systems of cultivation near Chapingo, Mexico. They found that disease levels were lower in the *chinampa* soils. When inoculum of *Pythium aphanidermatum* was introduced, the fungus was suppressed by *chinampa* soils. From their studies they concluded:

> In the chinampa agroecosystem, apparently a dynamic biological equilibrium exists in which intense management, especially of copious quantities of organic matter, maintains an elevated supply of organic nutrients and calcium, potassium and other mineral nutrients which stimulate biological activity in the soil. The elevated biological activity , especially of known antagonists such as *Trichoderma* spp., *Pseudomonas* spp., and *Fusarium* spp., can suppress the activity of *P. aphanidermatum*, other *Pythium* spp. and perhaps other soilborne plant pathogens.

Zuckerman et al. (1989), in a cooperative study between scientists from Mexico and the US, also studied suppression in *chinampa* soils, but of parasitic nematodes rather than fungi. The authors pointed out that the high organic content of the soil is probably responsible in part for the relatively few nematodes found, but nine organisms with antinematodal activity were also found. Their results were summarized as follows:

> Soil from the Chinampa agricultural system in the Valley of Mexico suppressed damage by plant parasitic nematodes in greenhouse and growth chamber trials. Sterilization of the chinampa soils resulted in a loss of the suppressive effect, thereby indicating that one or more biotic factors were responsible for the low incidence of nematode damage. Nine organisms were isolated from chinampa soil which showed antinematodal properties in culture. Naturally occurring populations of plant-parasitic nematodes were of lower incidence in chinampa soils than in Chapingo soil.

The *chinampas* are examples of traditional agricultural systems that used copious quantities of organic matter; thereby benefitting from natural biological control. Chapters 3 and 13 give additional details on the use of organic amendments and biological control.

Waru Waru

The story of the reconstruction of raised beds originally made by pre-Incan civilizations around Lake Titicaca has received widespread publicity. Erickson (1985), Erickson and Candler (1989), Garatcochea (1985, 1987), Lennon (1982), National Research Council (1989b), and Sattaur (1988) have described over 80,000 ha of raised fields or raised beds called *waru waru* or *caballónes* at an elevation of 3800 m near Lake Titicaca bordering Peru and Bolivia. Denevan (1985) suggested that some may be 2,000 years old. When a few of the raised beds were rebuilt according to specifications obtained from Erickson's archeological studies, potato yields in the 1983-84 crop season were 15 tons/ha from *caballónes*, while the regional average was 4.8 t/ha (Garatcochea 1987). Erickson and Candler (1989) reported potato yields for 1983-1986 of 8-14 tons/ha with an average of 10 tons/ha. Average potato yields in the Department of Puno were only 1-4 tons/ha. More recently, Straughan (1991) described yields of 20 t/ha from potatoes on raised beds reconstructed near Lake Titicaca in Bolivia. Yields were seven times those in nearby altiplano fields. In addition to better water management, the raised beds contributed to flood and frost control. Fish thrived in the canals between raised beds and ducks were also raised. The Bolivian and Peruvian governments now help farmers to reconstruct raised beds and some 50 ha have been reconstructed in Peru. Thus, farmers in the Andes are learning a forgotten technology from their traditional ancestors.

Tablones

Besides the *chinampas*, many other examples of raised fields are found today. The Maya in the Panajachel River delta of the highland Lake Atitlan basin, Guatemala, grow a large number of crops, especially vegetables, on raised beds called *tablones*, described by Mathewson (1984) and Wilken (1987). Both authors suggested that the *tablones* are probably pre-Colombian. Separated by irrigation trenches, they varied in height from 20-65 cm. During construction, a trench was made in the center of the *tablón*. Weeds, ground litter from nearby coffee plantations, and animal manure were placed in the trench, covered, and allowed to decompose. Later, muck from the irrigation ditches was sometimes added to the *tablón*. No mention was made of the possible presence or absence of root pathogens in this system, but the matter is worth investigating.

Flooding and Raised Beds

In Southern China, after two or three crops of paddy rice are grown in a field, the land is often ridged for growing various vegetables and ginger (King 1926, Williams 1981). A similar system was described in Taiwan (Su 1979). The anaerobic conditions prevalent under flooding for rice culture destroy many soilborne pests and pathogens (Cook and Baker 1983). A similar system is used in West and Central Africa. According to the International Center for Tropical Agriculture (IITA 1988), there are 85 million ha of inland valleys in sub-Saharan Africa, and 80% of the inland valley fields found in this region practice an annual cycle of mounding for vegetables, cassava, or sweet potatoes during the dry season and flat tillage for rice in the wet season. In the construction and destruction of mounds, organic matter and soil nutrients are recycled through incorporation of crop residues and weeds, and through flooding, many pests and pathogens in the soil are destroyed. Miracle (1967) also described the planting of rice in valley bottoms by the Tabwa people of Zaire. After the rice crop was harvested, mounds were made, which were planted to sweet potatoes or maize. Subsequently the mounds were flattened and rice was grown again. It is noteworthy that traditional farmers in widely separated areas of Asia and Africa developed similar systems that included flooding and raised beds, and these systems appeared to be effective against soilborne organisms. Chapter 11 discusses the effect of flooding on plant pathogens in greater detail.

Raised Beds in Asia

Raised beds, ridges, and mounds are commonly used in Asia for agriculture, especially in high rainfall and swampy areas (Chandler 1981, Harwood and Plucknett 1981, Herklots 1972, King 1926, Marten and Vityakon 1986, Milsum and Grist 1941, Villareal 1980). The importance of raised beds in traditional Asian agriculture is described by Herklots (1972) as follows:

> Throughout the monsoon countries of South-east Asia this is the almost universal technique adopted where water is abundant or too abundant. In the vast silted plains of the river valleys of Thailand, of the Mekong delta of Cambodia and of the Pearl River of South China and even of the smaller rivers of Taiwan raised vegetable beds are a feature of the countryside, especially near the large cities.

Hsu (1980) discussed the introduction to the Han Court in China of ridge farming (tai-t'ien) over 2,000 years ago. In many areas of China the use of raised beds is still common (FAO 1980). Ruddle and Zhong (1988) describe the "dike-pond system" found in the Pearl River Delta of Southern China. This intensive agriculture-aquaculture system has been evolving for the last two millennia and includes fish ponds and mulberry and sugar cane dikes. The cultivated land on the dikes also produces fruit trees, vegetables, and ornamental plants and flowers. The system includes an area of 800 square kilometers and supports a dense population of 1.2 million. Luo and Han (1990) call this raised bed system the "deep ditch-high bed system" and note that the ditches produce rice, water hyacinth, snails, and fish while the beds are intercropped with various vegetables, flowers, and fruit. There are about 48,000 ha of such raised beds in the Pearl River Delta. This integrated system provides an outstanding example of long-term agricultural sustainability.

Milsum and Grist (1941) described the formation of raised beds by Chinese gardeners in Malaysia. Raised beds in moist situations were as much as 0.6 m in height, but in dryer sites they were lower. Rotations were common; they noted that farmers rarely allowed a bed to have more than two successive crops. Extensive vegetable and flower growing on large raised beds is common in swampy areas near Bangkok, Thailand (Figure 14.3). Large quantities of organic material, consisting of manure, aquatic plants, mud from canals, and crop and plant debris, were incorporated into the soil in many Asian raised bed systems (FAO 1980, King 1926, McCalla and Plucknett 1981). The combination of raised beds, flooding, and copious quantities of organic matter added to the soil have contributed to better disease management and a sustainable agriculture.

Serpenti (1977) described the "garden islands" of the Kimain people of Frederik Hendrik Island, West Irian. Developed in swamps, these were raised beds similar to the Mexican *chinampas*. Regarding the use of the beds, Serpenti wrote: "Experience has taught the people exactly what the ground-water level should be at every stage of growth of every crop. Time of planting is chosen so that the changes in the ground-water levels combine most favorably with the level of the beds." Organic matter, consisting of grass, earth with high humus content, and ashes of sago palm leaves, was incorporated into the beds.

The Intha people living on Lake Inle in Burma and raise a variety of vegetables on floating islands made of grasses and reeds and fertilized with muck from the lake bottom (Jeffrey 1974). Peruvian Indians occasionally grow potatoes in reed boats on Lake Titicaca (personal communication -- John S. Niederhauser).

Root and Tuber Crops

An early account of the planting of cassava on mounds was that of Gonzalo Fernando de Oviedo (1986), who described the cassava planting practices of Caribbean Indians in 1526. The mounds they made were 1.8 m in circumference and "knee high". Six to ten stem pieces were planted per mound. Yams and sweet potatoes were also planted in mounds. Many authorities have reported that root and tuber crops such as cassava, yams, sweet potatoes, and taro were usually grown by traditional farmers in Asia, Africa, and the Americas in raised beds, mounds, and ridges (Barrau 1958, Coursey 1967, Denevan and Treacy 1987, IITA 1988, McLoughlin 1970, Miracle 1967, Nyoka 1983, Okafor and Fernandez 1987, Okigbo and Greenland 1976, Omohundro 1985, Waddell 1972, Wang 1983, Yen 1974a). According to Curwen and Hatt (1953), when Captain James Cook "discovered" New Caldonia in 1774, he found taro planted in ridges. Yen (1974a) reported taro growing on mounds in New Caldonia in 1965. Taro is found today planted on ridges or raised beds in most areas of Asia (Harwood and Plucknett 1981, Wang 1983, Yen 1974a).

Jones (1959) states that as early as 1854 the missionary David Livingstone saw cassava planted in raised oblong beds near the Belgian Congo. At this time Europeans thought cassava was indigenous to Africa whereas, in reality, it had been introduced into Africa with the slave trade by the Portuguese (Jones 1959). In Africa almost all yams are planted on mounds, ridges, or raised beds (Coursey 1967). Probably the most complete description of the use of mounds and raised beds by traditional African farmers is that of Miracle (1967). Organic material was commonly incorporated into mounds and raised beds. In several cases described by Miracle, organic matter placed in mounds was burned. Fresco (1986) mentioned burning organic matter incorporated into mounds for cassava in Zaire. Mounds and raised beds were used not only for root and tubers crops but also for vegetables, maize, and a variety of food crops. Prinz and Rauch (1987) described raised beds used in West Cameroon, which were adapted from a form of traditional mounds. Hahn et al. (1987) wrote that in West Africa, except where yams are grown on flat ground, they are usually planted on hand-made mounds, which may be 1 to 2 m in height and 2 to 3 m wide. Parsons and Denevan (1967) noted that many of the ancient South American raised fields probably were used for producing cassava, although positive evidence is lacking. Furthermore, Denevan and Turner (1974) suggested that root crops probably dominated raised bed agriculture in South America. Ibn al-Awam (1988), a Spanish Arab writer of the twelfth century, recommended that *caballónes* (raised beds) be made for radishes and onions.

In his fascinating ethnobotanical treatment of the sweet potato Yen (1974a) stated: "Within the range of methods of physical preparation of the soil for sweet potato cultivation, there is one physical effect which is universal -- the elevation of soil surface above normal for the area." Ridges, mounds, and raised beds are used for most of the sweet potatoes grown by traditional farmers. Yen (1974a) provided several examples of sweet potatoes growing on mounds, ridges, or raised beds. The Ifugao near Bontoc in the Philippines grew sweet potatoes in rotation with rice on circular mounds in their irrigated terraces.

The "mound builders" of New Guinea (Barrau 1958, Brass 1941, Lampert 1967, Yen 1974a) provide another example of traditional farmers who have developed a sustainable system of agriculture by cultivating sweet potatoes on mounds, producing high yields for long periods of time, with no apparent disease problems. Waddell (1972) described one of the systems in detail. Although the sweet potato gardens covered almost two-thirds of the cultivated land in his study area , other types of farming were practiced, such as mixed gardens, kitchen gardens, and cash-crop gardens. Sweet potato culture was primarily on large mounds called "*modó*." The mounds averaged 3.8 m in diameter and 0.6 m high. Smaller mounds were also made. The *modó* mounds permitted continuous cultivation without fallow. Sites in the study area were known to have been in continuous cultivation at least since 1938 (when Europeans first encountered these people). When a new mound was prepared, approximately 20 kg of old sweet potato vines, sugar cane leaves, and other sources of vegetation were placed in the center. When this material began to decompose, the mound was closed with soil and subsequently planted with sweet potato cuttings. According to Waddell (1972), the two to three harvests obtained per year totaled 19 tons/ha of sweet potato roots. The only reference to disease in this excellent, detailed study is the following : "It (sweet potato) is also less susceptible to disease than taro (*Colocasia esculenta*), which has suffered greatly in recent years from the depredation of the taro beetle (*Papuana* spp.) and the **virus** *Phytophthora colocasiae* in various parts of the Pacific." The above error regarding the nature of *Phytophthora* (an oömycete fungus -

not a virus) perhaps illustrates the level of knowledge and interest that most anthropologists, archaeologists, economists, ethnobotanists, geographers, and sociologists have developed regarding disease problems. Diseases are seldom mentioned or considered in their published studies of traditional or indigenous agriculture. One should not be unduly critical, however, since few plant pathologists study or cite work in those disciplines.

In their recently published compendium on sweet potato diseases, Clark and Moyer (1988) stated: "The rows may be prepared as level beds or raised beds, depending on local requirements. Raised beds are generally preferable to flat beds when it is necessary to improve drainage." Although most sweet potatoes in the US are grown on raised beds (personal communication -- C.A. Clark), the compendium did not mention the effect of raised beds on plant disease.

Maize Mounds and Ridges

The use of hilling, mounds, and ridges in maize culture appears to be an ancient practice in the Americas (Barrerio 1989, Weatherwax 1954, Wilson 1987). It was common throughout these continents. Clavigero (1974) stated in the 1700s that maize was hilled in Mexico because "it is better nourished and can resist sudden winds." When plants have reached a height of about 60 cm, a considerable amount of soil is mounded up around the base of the plant. In Guatemala, the practice is called *calzando*, "putting boots" on the corn. This practice is still common in higher elevations of Mexico, Central America, and parts of the Andes. Estrella (1986) reported that maize was planted with beans in mounds (*caballónes*) in Ecuador in 1573.

According to Carrier (1923) and Mt. Pleasant (1989), North American Indians, such as the Iroquois nation, also planted maize in mounds. The soil in the mounds was loosened at the time of planting. As the maize grew, soil was scraped around the roots and the mounds were carefully weeded. Crop residues and weeds were returned to the mounds, thus providing organic matter for the maize. Mounds were used repeatedly and thus became rather large. Remnants of such large mounds in abandoned fields were frequently found by early colonists in North America. Ridged fields, which had probably been used by Indians for maize culture, have been found in Mississippi, Georgia, Texas, Illinois, Michigan, and Wisconsin (Fowler 1969).

Reichel-Dolmatoff (1965) noted that in many parts of the San Jorge Valley in Northern Colombia one can observe hundreds of hectares covered with parallel ridges separated by furrows, the remnants of ancient raised beds used for maize and other crops. Whether these practices of growing maize in hills and ridges and on mounds had any effect on plant disease is unknown, but the better physical conditions for roots and better drainage probably reduced soilborne and crown diseases.

Knight (1978) described a system called "nkule" used in the grasslands of Tanzania. Grass was collected in piles, and soil was hoed on top of it. The grass under the mound was subsequently burned. A similar practice is used in Zaire (personal communication -- Diane Florini). Maize and cucurbits were then planted on the mounds. Fresco (1986) and Miracle (1967) described the use of like practices in the Congo Basin.

Some mound systems are rather complicated and ingenious in that they make use of existing resources in a treeless environment and take advantage of nutrients provided by decomposing grass and, after the mounds have been formed, leguminous plants. Strømgaard (1988) described a system used by the Aisa-Mambwe in Zambia illustrative of this complexity. First, circular mounds (fundikila) were made by cutting up grass sod and piling it into mounds with the turf side inwards.

> The compost-mounds, *fundikila*, made towards the end of the rainy season, are either left to rot for the remainder of the dry season or grown with beans and a little cassava in the mounded 1-year garden, *ntumba*. Next year the mounds are thrown down, and in the garden, *icalo*, maize and later millet are broadcast between the growing cassava of the first year. At the start of the following rainy season, in November, the Mambwe normally throw the garden up into mounds for beans and groundnuts with some millet and a little cassava grown between the mounds; in this stage the garden is called *ichitikula*. If mounds are not made, maize and cassava are grown on the flat, *mpepe*. The *ichitkula* is followed by *mpepe*, as above, after which the peasant either decided to throw up the garden into mounds for the third time, *ichituklulu*, with beans, or continue the flat *mpepe*, this time with groundnuts.

Raised Beds for Plant Disease Management

In addition to the obvious benefits of better water management, raised beds, ridges, mounds, and hills are undoubtedly also used because of their value in reducing a high incidence of various root rots in poorly drained soils. Many investigators have found that flooding and associated oxygen stress predispose or increase the susceptibility of plants to infection by various species of *Phytophthora*, *Pythium*, and *Aphanomyces* (Barta and Schmitthenner 1986, Cook and Baker 1983, Kuan and Erwin 1980, Matheron and Mircetich 1985, Mueller and Fick 1987, Stolzy et al. 1965, Wicks and Lee 1985, Wilcox and Miretich 1985). The literature on plant water stress as a predisposing factor to plant disease has been reviewed by Ayres and Boddy (1986), Cook and Baker (1983), Kozlowski (1978), Schoeneweiss (1975, 1986), and Stolzy and Sojka (1984). Raised beds would often prevent or reduce such predisposition to disease due to flooding.

Several species of *Phytophthora* cause serious root rots of cassava in tropical areas (Booth 1977, Oliveros et al. 1974). Planting in well-drained raised beds or ridges was found to be an effective practice for reducing cassava root rots caused by *Phytophthora* spp. (Booth 1977, Lozano and Terry 1976). Yields of cassava planted near Caicedonia, Colombia without ridges were reduced to 7 tons/ha. by *Phytophthora* spp. After farmers planted in ridges, following CIAT recommendations, cassava yields rose to an average of 23 tons/ha. in a 20,000 ha area (personal communication -- J. C. Lozano).

When cow dung was added to yam mounds before planting in Ghana yields were increased and nematode numbers (*Scutellonema bradys*) were significantly reduced, according to Adesiyan and Adeniji (1976).

For potato production, exceptionally large hills (0.7-0.9 m high) were commonly made by traditional farmers in some parts of the Andes Mountains. I have seen such large hills in several countries in the Andes. Padre Bernabé Cobo (Mateos 1956) wrote in the seventeenth century that Peruvian Indians preparing land with the Andean foot plow (the taclla) made *caballónes* (large hills); he added "y muy grande algunos" (that some hills were very large). These large hills were made for planting potatoes. In my experience, tuber infection caused by the fungus *Phytophthora infestans* (causal agent of late blight of potatoes) was rare in the Colombian Andes. The soil of the large hills probably filtered out the fungal spores before they could reach the tubers (Thurston and Schultz 1981). Goodall et al. (1987) and Coffey (1984) found that mounds aided in the management of *Phytophthora cinnamomi*, a root rot pathogen of avocado trees. Thus, planting in large hills, mounds, and ridges appears to contribute significantly to disease management .

Some examples of the contribution of raised beds or ridges to the management of pathogens are found in the more modern literature. The losses caused by *Erwinia caratovora*, causal agent of bacterial softrot of Chinese cabbage, is reduced by the use of raised beds or ridges in the US (Fritz and Honma 1987) and China (Williams 1981). Ridging is recommended for the management of bottom rot of lettuce (*Rhizoctonia* spp.) in New York (Pieczarka and Lorbeer 1974). Heart rot and root rot of pineapple (*Phytophthora nicotianae* var. *parasitica*) was reduced in raised beds, according to Baker (1938) as cited by Shea and Broadbent (1983). According to Abawi (1989), in Popayan, Colombia, *Rhizoctonia solani* is less severe during the wet rainy season if beans are planted on raised beds, which facilitate good drainage. Abawi and Pastor-Corrales (1990) wrote that growing beans on raised beds or ridges reduced diseases that are favored by high soil moisture, such as southern blight (*Sclerotium rolfsii*), Rhizoctonia root rot (*Rhizoctonia solani*), and Pythium root rot (*Pythium* spp.). Raised beds are used extensively in California and contribute to the management of the red stele disease of strawberry (*Phytophthora fragariae*) and various root rots of lettuce. Raised beds are recommended for management of leather rot of strawberries (*Phytophthora cactorum*) in Ohio (Madden et al. 1919). Knowles and Miller (1965) recommend that safflower be planted in raised beds in California for the management of *Phytophthora drechsleri*. Johnston and Springer (1977) found that planting in ridges in New Jersey contributed to the management of *Phytophthora capsici* (*Phytophthora* blight of pepper). Arneson (1971) found that raised beds helped to manage white rot of peanut (caused by *Sclerotium rolfsii*) in Nicaragua. Abawi, Crosier, and Cobb (1985) recommended raised beds for management of *Pythium* root rot of snap beans in New York. Despite such reports, raised field technologies receive surprisingly little if any attention in most plant pathology texts, and the terms "raised beds" and "ridges" are almost never found in the indices of books about plant pathology.

Summary

Raised fields, raised beds, ridges, and mounds were used widely for millennia by traditional farmers in geographically separated areas of tropical America, Asia, and Africa. Raised bed systems of agriculture with striking similarities evolved in these areas. Drainage, fertilization, frost control, and irrigation were among important considerations in these systems, but planting in soil raised above the soil surface is also an important disease management practice for soilborne pathogens. I have described specialized raised bed systems such as *chinampas*, *tablones*, and *waru waru*. Almost all of the systems included the incorporation of organic matter into the raised beds. How much the management of plant diseases and other pests entered into the evolution of these systems is unknown. Raised beds are used extensively today in Asia, often after a rice crop, as flooding the soil for rice culture destroys many soilborne pests and pathogens, and subsequently vegetables and other crops can be grown on the semi-sterilized soil in the raised beds with fewer disease problems. Similar practices are used after rice in tropical Africa. Today, mounds, ridges, and raised beds are used worldwide by indigenous farmers for root and tuber crops, and their use reduces root rot problems. I have listed a number of the diseases managed by raised beds in modern agriculture. The widespread use of raised beds in agriculture today testifies to their value.

Most knowledge regarding raised beds in traditional agriculture comes from anthropologists, archaeologists, ecologists, geographers, and ethnobotanists rather than from plant pathologists or other agricultural scientists. Cooperation and communication among disciplines to make the principles and knowledge regarding the merits of raised beds available to farmers worldwide would be highly desirable. Some of the raised bed systems, such as the *chinampas*, are rapidly disappearing (due to urban encroachment). Establishing "natural preserves" where such traditional systems can be maintained would be a contribution to the enlightenment and well-being of future generations. Determining the value of these practices should become part of the research agenda of governments and international agencies in developing countries.

15

Rotations

Crop rotation is an ancient agricultural practice that, in addition to its agronomic value, is important in managing certain plant pathogens, especially those in the soil. The slash and burn agricultural system includes both fallow and rotation of fields and has been used for millennia. The Chinese have used rotations for thousands of years (FAO 1980, Wittwer et al. 1987). The Romans used legumes such as alfalfa, peas, broad beans, vetches, and lupines in their crop rotations and had different rotational schemes for different types of soils (Cato 1934, Columela 1988, Spurr 1986, Watson 1983, White 1970). Cato (1934), a Roman who lived in 234-149 B.C., wrote that lupines, beans, and vetch fertilized the land.

The Roman Virgil (70-19 B.C.), as translated by Lewis (1941) recommended the following:

> See, too, that your arable land lies fallow in due rotation, and leave the idle field alone to recoup its strength: or else, changing the seasons, put down to yellow spelt, a field where before you raised the bean with its rattling pods, or the small seeded vetch, or the brittle stalks and rustling haulm of the bitter lupine, so too are the fields rested by a rotation of crops, and unploughed land promises to pay you.

In Almeria, Spain, during the thirteenth century (Equaras Ibañez 1988) the Moor Ibn Luyun observed: "You should not plant wheat or millet more that two times in succession in the same field, because the soil has an aversion to repeated plantings with the same cereal." The diverse rotation and fallow sequences used today by traditional farmers in Bolivia are described in detail by Hatch (1983).

Nature of Rotation

Curl (1963) defined rotation as follows: "the growing of economic plants in recurring succession and in definite sequence on the same land as distinguished from a one-crop system." The National Research Council (1989a) stated that, "crop rotation is the successive planting of different crops in the same field." Rotations often include fallow periods at periodic intervals. The term "rotation" will be used throughout this book, although some authors prefer the term "crop sequence" as rotation may imply sowing of the same crop at regular intervals. The literature on rotation is voluminous, and no attempt will be made to provide a comprehensive review. However, conclusions can be drawn from the literature that effective use of rotation, relative to the management of plant diseases, seems to be highly specific, according to location, crop, and pathogen.

Palti (1981) gives the following reasons for rotation: (1) more efficient use of nutrients, (2) improved soil texture, (3) water conservation, (4) weed control, (5) management of soilborne pathogens. Palti further notes that advances in modern agriculture have made some of the agronomic reasons for rotation less important. Weeds can be efficiently controlled with herbicides, fertilizers have become economic, and soil texture can be maintained with proper tillage and soil amendments. As the costs of pesticides, fertilizers, and energy increase, some of Palti's conclusions may not hold for modern agriculture, and certainly are not relevant to poor, traditional farmers. However, water conservation and disease management are relevant to almost all farmers, both modern and traditional.

Rotations and Plant Disease

There are literally hundreds of references on the use of rotations in plant disease management. Curl's (1963) article on the control of plant diseases by rotation included 415 references and is probably the most comprehensive review of the subject. Many other authors (Baker and Cook 1974, Cook and Baker 1983, Garrett 1970, Glynne 1965, Good 1968, Leighty 1938, Luc et al. 1990, Palti 1981, Stakman and Harrar 1957, Stevens 1960, Sumner et al. 1981, Walker 1950) have also considered crop rotation relative to plant disease.

The value of rotations for the management of fungal pathogens depends in part on the nature of the pathogen. Zadoks and Schein (1979) noted that if a pathogen attacks many host species rotation may not be successful, whereas if a pathogen attacks only one host species rotation is more likely to succeed. Rotation is less effective against wind-borne pathogens than rain-splash disseminated pathogens or slow spreading soilborne pathogens. Also, if the pathogen can maintain itself on plant debris in the soil, rotation may be less successful.

Many plant pathogens cannot survive extended periods in the soil, and this "starving out" is one of the several values of crop rotation. Crop rotation is basic to the management of root diseases. By growing different crops with differential susceptibility to root pathogens, those pathogens that do not have the capacity to survive long periods in the absence of a host can often be managed successfully. In addition, some pathogens whose propagules are primarily airborne may become more destructive in monocultures due to the buildup of overseasoning inoculum. Some cereal smuts, late and early blight of potatoes (*Phytophthora infestans* and *Alternaria solani*), Sigatoka of bananas (*Mycosphaerella musicola*), and many viral diseases (in their vector infected hosts) are examples of such diseases (Stakman and Harrar 1957).

Pathogens can seldom be completely eradicated by rotations, thus disease management rather than complete control is usually the objective of crop rotation. Crop rotation is often used in conjunction with other cultural practices such as the use of organic amendments, fallow periods, sanitation practices, and manipulations of moisture and nutrients. Indirectly, rotation is often a form of biological control, as it influences the microbial activity of the soil..

Chinese farmers have used rotations for millennia. Their rotational schemes are often complex (FAO 1980, Wittwer et al. 1987). Williams (1981) stated that in intensively cropped vegetable areas in China, "farmers are acutely aware of the need to rotate various types of vegetables for the prevention of soilborne diseases and pests." Their rotations take into account the danger of consecutive cropping of vegetables susceptible to the same pathogen. The use of flooding in rotation schemes also plays an important role in the suppression of soilborne pathogens.

Granados-A. et al. (1990) suggested that in the state of Tabasco, Mexico, traditional farmers evolved rotation systems using legumes such as *Canavalia* spp., *Pueraria* spp. and *Stizolobium* spp., that reduced losses caused by soilborne diseases. The rotations permitted a sustainable corn agroecosystem, that produced 3-4 t/ha of maize. *Stizolobium* spp., locally know as "*nescafe*", was the most effective plant in rotations for maintaining fertility and is now planted on more than 4,600 ha in that state. Arevelo-R. and Jimenez (1988) also discussed the extensive use of velvet bean (*Stizolobium pruriens* var. *utilis*) in Uxpanapa, Mexico.

Rotations and Nematode Management

Rotations are an important tool in nematode management. Good (1968), Murphy et al. (1974), and Nusbaum and Ferris (1973) provided useful reviews on the subject. Recommendations for nematode management in tropical and subtropical crops were reviewed by Luc et al. (1990). Good gave many examples of rotation for nematode management, but concluded: "Crop rotations and other soil management practices will not control all nematodes that injure crop plants because of the overlapping host susceptibility of cultivated and forage crops." For example, Crotalaria, millet, bermudagrass (cultivar "Costa"), bahiagrass, sudangrass, and marigold are crops effective in rotations against several species of *Meloidogyne* (root-knot nematodes). Unfortunately, these same crops are hosts to other parasitic nematode species (Murphy, et al. 1974). Thorne (1961) wrote that a 3-4 year rotation manages nematodes with a narrow host range, such as the sugar beet cyst nematode (*Heterodera schachtii*) or the alfalfa stem nematode (*Ditylenchus dipsaci*). However, crop rotation for the management of nematodes with a broad host range, such as *Meloidogyne* spp., has usually been found to be unsuccessful.

The root-knot nematodes (*Meloidogyne* spp.) are able to attack over 2,000 species of plants (Bird 1978) and are considered among the most important crop pathogens worldwide. It is often difficult to identify alternate crops for rotations, but Brodie (1984) found that a 99% reduction of root-knot nematode populations can be achieved in one year in some soils when no host is present. In tropical countries, where nematicides are unavailable or expensive, crop rotations, especially with grasses, can often provide effective management. Work in Puerto Rico (Ayala 1968) and Florida (Guzman et al. 1973) has shown that one year of rotation with pangola grass in sandy soils is sufficient to suppress *Meloidogyne incognita* populations. Without rotations, nematode populations often become destructive. This occurred in Hawaii where populations of *Rotylenchus reniformis* became a major problem on pineapple grown in monoculture for 35-40 years (Rohrbach and Apt 1986).

The slash and burn or shifting cultivation system can be considered a rotation system, involving a rotation of fields rather than crops. Usually, successive crops of the same species are harvested until the plot is left to fallow or abandoned because weeds have become unmanageable or soil fertility exhausted. Wilson and Caveness (1980) reported from Nigeria: "Plant parasitic nematodes are among the major soilborne plant pests that are effectively suppressed by the bush fallow (*slash and burn*) system and for which effective controls must be developed where alternatives to the bush fallow are being considered." In another study in Nigeria, Caveness (1972b) cleared and then prepared 19 plots representing five traditional farming systems. Different crops were planted and subsequent nematode populations recorded. From these studies he concluded: "These data suggest that under the traditional farming system of shifting cultivation the several plant parasitic nematode species are able to survive, albeit in small numbers, and that modern farming practices favor the increase of certain species of nematodes to the detriment of others." High nematode populations were found in continuous crop cultivation, while low populations were found in bush fallow rotations according to Nickel (1973). Jurion and Henry (1969) discussed the many different crop rotations used in tropical Africa.

The management of the destructive potato cyst nematodes (*Globodera pallida*; *G. rostochiensis*) by the Incas of Peru in the South American Andes gives another example of the value of rotation (in this case a 7-8 year rotation in combination with fallow) in nematode management. This example is described in detail in Chapter 9.

Although it may have little to do with nematode management, it is coincidental and interesting that by trial and error, the first President of the US, George Washington, worked out a 7 year crop rotation on his farm at Mount Vernon in the 1700s. The seventh year he planted potatoes and maize (Dies 1949).

Rotation Versus Monoculture

Shipton (1977) reviewed the literature on monoculture and soilborne plant pathogens. Monoculture has been defined as the practice of cultivation of the same crop in the same soil year after year (Kupers 1972). Shipton suggested that rotations have been common in European agriculture since the Middle Ages, while monocultures have been more common in the US. Monocultures are generally thought to increase plant disease and result in a steady decline in yields due to the buildup of soilborne pathogens. However, monocultures do not necessarily always lead to increased disease. Shipton (1977) identified two patterns of disease occurrence during monoculture. In the first or "irreversible" pattern, the disease incidence tends to become constant in some soil pathogen/host combinations. In the other, "reversible" pattern, disease develops, but then tends to decline over an extended period of time as suppressive soils evolve. The take-all disease of wheat

(*Gaeumannomyces graminis* var. *tritici*) is commonly used as an example of reversible disease decline. According to Baker and Cook (1974), take-all severity generally increases for 2-4 years under wheat monoculture and then declines in the following years of wheat monoculture. Palti (1981) gave additional examples of reversible and irreversible diseases.

In the tropics some crops are usually grown in monocultures for extended periods. According to Ruthenberg (1980), rice has been grown annually on terraces in the highlands of Luzon for over 2,000 years, apparently without serious disease problems. However, no record of disease problems during this time was made. Bananas, sugar cane, sisal, coconuts, oil palm, and pepper are further examples of tropical crops grown in monocultures for extended periods. In Uganda the Ganda people maintained plantains in the same field for up to 50 years without rotation, by careful pruning, weeding, and the use of mulches (Fallers 1960). Tree and shrub crops such as coffee, cacao, rubber, tea, citrus, and other fruit crops can also be considered examples of tropical monocultures, but they often have serious problems with soilborne pathogens (Fox 1970).

Rosado May et al. (1985) studied five different crop rotations in Tabasco, Mexico, investigating soil populations of *Rhizoctonia solani*, *Pythium* spp. and *Fusarium* spp., in the systems. They found a reduced incidence of these organisms and less crop loss in soils under a bean and maize rotation as compared to a maize monoculture. Low organic matter content of the soil seemed to directly influence the high incidence of soilborne plant pathogens in their studies. Traditional farmers in Mexico and Central America use high levels of organic matter whenever possible in their agriculture (Wilken 1987).

Perhaps the oldest experiments on monoculture and rotation are the Broadbalk wheat experiments begun at Rothamsted, England, in 1843 (Glynne 1965, Shipton 1977). Wheat has been grown continuously there for almost 150 years. Long-term experiments showed that cereals tend to come to an equilibrium with soilborne plant pathogens in a monoculture. Additional examples of successful monoculture could be given (Shipton 1977), but many failures due to disease could be cited also. Shipton (1977) analyzed the effect of monoculture on plant diseases. Knowledge of the environment, the cropping system, the ecosystem in which a crop is grown, and the etiology and epidemiology of existing pathogens becomes essential to the design of effective management strategies utilizing rotations.

Summary

Rotation is an important and ancient practice in traditional agriculture. Rotations contribute to the management of both soilborne and airborne pathogens, but also provide other benefits such as improved soil texture and better use of nutrients and water. There are hundreds of references to rotation in ancient writings on agriculture, and rotations are one of the best documented traditional agriculture practices.

Modern agriculture in developed countries has raised concern because it is highly energy-intensive, has a narrow genetic base, and its emphasis on increasingly high yields and efficiency relies too much on monocultures. Sometimes destructive erosion, pollution, and excessive pesticide residues result. However, properly managed, monocultures do not necessarily produce negative effects. Extensive monoculture of tropical crops important to traditional farmers, such as plantains, coconuts, sisal, oil palm, coffee, cacao, and citrus, attest to this fact.

Rotation is especially important in the management of soilborne plant pathogens. There is a voluminous literature on the use of rotations in plant disease management, but little was found specifically on rotation as related to plant disease management in traditional systems. Even the extensive literature on the slash and burn system, which involves rotation and fallow, and which is still important in many tropical regions, has few references regarding the effects of slash and burn rotations on plant diseases. Rotations are especially important in nematode management, and there is evidence that ancient farmers, as well as today's traditional farmers, utilized rotations for nematode management.

The value of rotations for disease management is highly location specific; nevertheless, the use of rotations should be carefully studied and, where feasible, utilized in schemes designed to aid traditional farmers. Well-planned research on the long-term effects of rotation in the most important traditional cropping systems would undoubtedly have important payoffs in the future.

16

Terraces

Terraces for agriculture are found in many mountainous areas of the world. Because their construction and use is highly labor intensive and usually precludes mechanization, terraces today are used primarily by traditional farmers. Some of the practices used in constructing and maintaining terraces contribute to the management of plant diseases. Specifically, I suggest that the incorporation of organic matter into terrace soils contributes in a major way to assuring their continual use for long periods of time.

Terraces reduce erosion, manage irrigation water, and provide a level surface on which to cultivate crops. The retaining walls of terraces for agriculture are constructed of various materials, mainly stone, on sloping land in mountainous areas. Terracing by the use of vegetative strips is becoming of increasing importance There are innumerable variations of types and subtypes of terraces (Bunch 1982, Denevan 1987, Donkin 1979, Spencer and Hale 1961, Webster and Wilson 1980, Wilken 1987). According to Wilken (1987), some bench terraces on hillsides are shaped with hoes and may be later abandoned, while other terraces are carefully maintained and improved over long periods of time. He described in detail the construction of various types of terraces with various materials in Mexico and Central America.

Terraces have been used extensively in Africa (Miracle 1967, Netting 1968, Mohamed and Teri 1989) and Asia (Conklin 1980, Geertz 1963, Wittwer et al. 1987). Han (1987b) related that more than half of China's arable land is on mountain slopes and, although figures were not given, large areas in China are covered by terraces. Mohamed and Teri (1989) stated that all farmers in the Mgeta Division of Tanzania used terraces, and that they incorporated crop residues and weeds into them.

Terraces in Latin America

Traditional peoples in the Americas have used terraces extensively (Denevan 1985, 1987, Donkin 1979, Mateos 1956, Puleston 1978, Ravines 1978, Schjellerup, 1985, Turner 1978b, Wilken 1987). For example, Denevan (1985) related that there are about one million hectares of terraces in Peru, but that over half of them are abandoned. The total arable land in Peru today is only 2.4 million ha. Schjellerup (1985) described in detail a number of different ancient terracing systems found in Northern Peru. Almost all of the Peruvian terraces were built in pre-Columbian times. The question of why terraces were abandoned is a critical one, since recommendations are being made in many developing countries, not only for the renovation of old terraces, but for the construction of new terraces.

The ancient Maya Indians constructed large areas of terraces in Belize, and according to Turner (1974), "tens of thousands of relic terraces crisscross the hillsides of Southern Campeche and Quintana Roo" in Mexico. Puleston (1978) and Wilken (1971) reviewed the literature on Mayan terraces in Mexico and Central America.

Cook (1916) described the construction of the ancient Peruvian terraces. All of the terraces investigated had the same internal structure. The terraces consisted of an outside wall and two distinct layers of soil behind the wall. The lower layer was composed of stones and clay. It was covered by a layer of fine agricultural soil 0.6-0.9 m thick. The terraces Cook described on steep slopes were only 0.9-1.2 m wide, whereas those on less steep land were 2.4-4.6 m wide. They were usually 2.4-4.3 m high. Irrigation of the terraces was accomplished by artificial channels leading down from water sources such as glaciers in the high mountains (Ravines 1978, 1980). These feats of hydraulic engineering were described as among the most outstanding in the world. Weatherwax (1954) described and discussed Peruvian terraces and their use for maize cultivation. He related that human waste was highly regarded as a fertilizer for maize in Peru at the time of the Incas, and was dried, pulverized, and stored until the time for planting maize. Gade (1975) noted that the soil in the terraces may have been transported from valley bottoms, and that llama trains from the coast transported *guano* for fertilizer to the highlands during the time of the Incas.

There are numerous references to the extensive terracing systems found in Peru (Del Busto 1978, Denevan 1987, Gade 1975, Guillet 1987, Mateos 1956, Poma de Ayala 1987, Schjellerup 1985, Treacy 1989, von Hagen 1959). A study of the causes of terrace abandonment in Peru was the major objective of a project in the Colca Valley of Peru (Denevan 1987). Denevan noted that depopulation and climatic changes were the most common causes suggested for terrace abandonment. Other possible causes suggested for the breakdown of the agroecology of the regions after the Spanish conquest included the introduction of animals, which trampled and damaged terraces (Poma de Ayala 1987), and the Spanish destruction of the Inca administrative system, that organized and managed the highly complex terrace system. Denevan concluded that a combination of factors may have contributed to terrace abandonment. He described a number of different types of terraces in the Colca Valley including bench terraces, broadfield terraces, valley-bottom walled terraces, upland walled terraces, and sloping-field terraces. He noted that the soils of the terraces were "now maintained by the application of manure and compost and by periodic fallowing, and both were likely true prehistorically." Treacy (1989) reported that at the present time terraces in the Colca Valley are maintained with burro and sheep manure.

Mountjoy and Gliessman (1988) described a terrace/*cajete* system used by traditional farmers in Tlaxcala, Mexico. There is archaeological evidence of terraces as early as 1000 B.C. in the area. Cajetes (boxes in English) are channels divided into tanks excavated at the base of terraces. They reduce erosion and catch field runoff during rains, thus trapping eroded soil, organic litter, and other nutrients, and allow water to percolate slowly into the soil. The cajetes also act as compost pits. They are periodically emptied, and their contents are incorporated into the terrace fields.

In Guatemala, Mayan wheat farmers incorporated wheat stubble into terraces. Wilken (1987) stated:

> Rather than disturbing the previous season's surface, farmers shave soil from terrace risers with *azadones* (hoes) to cover old surfaces and stubble or ash. By burying old surfaces in this manner, farmers inter organic matter at the root levels of the new crops and reduce the potential for erosion because basic terraces structures remain undisturbed and reinforced with intact root systems.

He added:

> In most crop systems, weeds that have grown since the last cultivation are simply turned under along with whatever crop residues remain on the surface.

Terraces in Asia

Upawansa (1989) suggested that terraces were an efficient soil conservation measure in Sri Lanka. Soil that washed down from the highlands settled in terraced paddy fields and tanks. Upawansa pointed to the fact that small village tanks remained unsilted for several centuries in terraced areas, as evidence of the value of terraces against soil erosion.

Barker (1990) described an erosion control practice of the Ikalahan shifting cultivators of Northern Luzon, the Philippines, whose major crop is sweet potatoes. On steep hillsides when a plot of sweet potatoes was harvested, the tops and stems of the harvested plants were placed in a shallow trench dug on the contour, and covered with soil. The practice is called "gen gen." These "gen gen" strips soon became vegetative terraces effective in preventing erosion. The large quantities of composted organic matter in the trenches increased yield and probably played a positive role in reducing soilborne pathogen problems (see Chapter 13 on organic amendments). Very few soilborne pathogen problems were observed on sweet potatoes in "gen gen's" during a 3 and 1/2 year period of observation (personal communication -- T. C. Barker).

Yen (1974a) provided a fascinating account of sweet potato culture on mounds, ridges, raised beds, and terraces. In the Philippines, Ifugao farmers near Bontoc grow sweet potatoes on circular mounds in their irrigated terraces in rotation with rice. Yen noted that there may be a rice-rice succession or a rice-sweet potato succession. Organic matter is incorporated into the terraces:

> In the spring preparation of fields for rice planting, large parties cut and bring in selected shrubby second growth from nearby gullies for incorporation into the soil by puddling with hands and feet. The soil has been made ready for this by early precrop inundations. After rice harvest, and the drainage of any residual water, preparation for the sweet potato planting entails the incorporation of the cereal stubble.

The incorporation of such organic matter helps to explain why these terraces have been used continuously for over 2,000 years.

Summary

Perhaps terraces do not deserve a separate chapter in this book. However, the question of why it has been possible for traditional farmers to use terraces for centuries and in some cases for millennia is important, especially as recommendations are being made in many developing countries, not only for their renovation, but for the construction of new terraces.

Murra (1960) suggested that the terraces in Peru were used primarily for maize during the time of the Incas. Von Hagen (1959) quoted Cieza de León as describing Inca practices in the 1500s as follows: "They bring back the droppings (*guano*) of birds to fertilize their cornfields and gardens, and this greatly enriches the ground and increases its yield, even if it once was barren. If they fail to use this manure, they gather little corn." Although terraces were not specifically mentioned, it is virtually certain, because of the importance of maize to the Incas, that they received *guano*. Denevan (1987) writes that at the time of the arrival of the Spanish, the terraces in the Colca Valley of Peru were used for maize, quinoa, beans, and tuber crops. Without long rotations potatoes probably would not do well in terrace soil for extended periods of time, unless the organic matter added to terrace soil enhanced the soil's suppressiveness for soilborne pathogens. In Peru, before the arrival of the Spanish, the ruling Incas used fallow and rotations in the production of potatoes according to Garcilaso de la Vega (1966) and Poma de Ayala (1987). Long rotations of 6-8 years are found today in Peru (Brush 1980, Brush et al. 1981, Mayer and Fonseca 1979). Rotations were doubtless used for terraced land also.

Long-term experiments at Rothamsted, England, in which wheat had been grown continuously since 1843, showed that cereals come to an equilibrium with soilborne plant pathogens in a monoculture (Glynne 1965, Shipton 1977). Perhaps a similar equilibrium with soilborne pathogens occurred in the terrace soils. Mohamed and Teri (1989) stated that all farmers in the Mgeta Division of Tanzania used terraces and incorporated crop residues and weeds into them. Ruthenberg (1980) wrote that rice was grown yearly in the Ifugao rice terraces in Northern Luzon in the Philippines for 2,000 years without apparent serious problems. Rice alternates with sweet potatoes in the Ifugao terraces, and the crop residues from both crops are incorporated into the terrace soils. The drainage and water-retention properties of the soils in the terraces also affects populations of soilborne pathogens there. The incorporation of large quantities of organic matter into the soils of terraces helps to explain the sustainability of these systems over centuries.

17

Tillage

Tillage practices include plowing, chiseling, disking, harrowing, leveling, cultivating, ridging, and subsoiling. Ancient Africans, American Indians, Arabs, Chinese, Egyptians, Greeks, Indians, and Romans often had rather sophisticated agricultural systems that used many of these practices. According to Warren (1983) tillage has been used primarily for weed control, seedbed preparation, and "for some supposed value of loosening the soil." Plowing generally produces various degrees of soil inversion. Although the major value of plowing seems to be for weed control, destruction by burying of crop residues infested with the long-term survival structures of plant pathogens is also of considerable value and an important plant disease management practice.

Effects of Tillage on Plant Disease

Palti (1981) listed as follows the effects tillage can have that may influence plant disease:

1. Plowing incorporates various types of organic matter into the soil (crop residues, manure, green manure, volunteers, weeds).
2. Tillage buries plant pathogens from the topsoil into deeper layers of the soil where they cause less or no disease.
3. Tillage operations move soilborne pathogens from isolated foci into other areas of the field.
4. Tillage shapes the soil into hills, ridges, or raised beds, that provide better drainage and irrigation.
5. Successive tillage operations can reduce the inoculum of some pathogens by solarization.

6. Tillage reduces weed populations and volunteers that may play a role in disease.

Modern Minimum Tillage Systems

Minimum tillage has become important to modern agriculture in recent years. For example, the National Research Council (1989a) estimated that nearly 40 million hectares were farmed using some form of conservation tillage in the US in 1987, compared to only 16 million in 1980. Phillips et al. (1980) noted that the earliest systems of agriculture were often essentially minimum or no-tillage systems. Today, many systems of traditional agriculture are still minimum or no-tillage systems. A variety of terms are used for systems using less tillage than conventional tillage agriculture. No-tillage, minimal tillage, conservation tillage, ecofallow, limited tillage, minimum tillage, mulch tillage, reduced tillage, surface tillage, and zero tillage are only some of the terms found in today's literature (American Society of Agronomy 1978, Crosson 1981, Greenland 1975, Hayes 1982, Phillips and Phillips 1984, Rockwood and Lal 1974, Sprauge and Triplett 1986, Sumner et al. 1981, Young 1982). There are in the literature innumerable definitions of the above terms. The Soil Conservation Society of America (1976) defined no-tillage in this way: "A method of planting crops that involves no seedbed preparation other than opening the soil for the purpose of placing the seed at the intended depth. This usually involves opening a small slit or punching a hole into the soil. There is usually no cultivation during crop production. Chemical control is normally used."

Possible advantages of no-tillage systems were listed by Warren (1983):

1. can be used on steep or rocky land where animal or mechanized tillage is difficult or impossible
2. less energy required for crop production
3. reduced soil erosion by wind and water
4. improved management of soil moisture
5. may improve soil structure and soil organic matter content
6. reduced weed problems by mulch on soil surface
7. soil temperatures and daily temperature fluctuations in the soil are reduced
8. reduces injury to roots of crop plants caused by mechanical tillage or hand weeding
9. may reduce some insect problems
10. reduces spread of pathogens by tillage activities

11. may reduce some soilborne diseases by reducing tillage wounds, which facilitate entry of soilborne pathogens

Warren also listed possible disadvantages of no-tillage systems:

1. pest problems (insects, diseases, slugs, etc.) may be increased
2. serious weed problems result unless the systems effectively control them

Today's modern no-tillage or minimum tillage systems (American Society of Agronomy 1978, Hayes 1982, Phillips and Phillips 1984, Young 1982) depend heavily on the use of herbicides. Greenland (1975) suggested that herbicides offer "freedom from the hoe" for traditional farmers using zero tillage in the tropics, and he further suggested that an "ultra-low-volume" knapsack sprayer was a less expensive and more appropriate technology than tractors and mechanization for small farmers. The expense, possible toxicology, and environmental effects of herbicides should be carefully evaluated and considered before recommending their use by traditional farmers.

Effects of Crop Residues on Plant Diseases

Cook et al. (1978) and Boosalis et al. (1986) reviewed the literature on the effect of crop residues on plant disease. Cook et al. (1978) listed three ways in which crop residues left in the field (i.e. by no-tillage) can affect plant diseases:

1. For many plant pathogens, residues provide food and a place to live and reproduce
2. Residues affect the physical environment occupied by the host and pathogens
3. As organic soil amendments, residues intensify the microbial activity of the soil and this along with a variety of decomposition products (some phytotoxic or fungitoxic) may affect pathogens, susceptibility of the host plants, or both.

A number of fungal pathogens may survive in residues in no-tillage systems. For example, Sumner et al. (1981) listed six foliar pathogens of maize, that may survive in residues: *Cochliobolus heterostrophus*, *Setosphaeria turcica*, *Phyllosticta maydis*, *Physoderma maydis*, *Colletotrichum graminicola*, and *Cercospora zeae-maydis*. They added that most of these pathogens caused diseases more damaging in minimal tillage than in conventional tillage, where debris is plowed under. On the other hand, minimal tillage has decreased the importance of the take-all disease of wheat (*Gaeumannomyces graminis* var. *tritici*). Doupnik et al. (1975) reported that ecofallow reduced stalk rot in grain sorghum.

Sumner et al. (1981) reviewed the literature on the effect of no-tillage and plowing under crop debris on a number of other plant pathogens. They concluded that the effectiveness of plowing under crop residues on the subsequent development of a disease depends on the epidemiology of the pathogen. The practice is highly effective with some fungal pathogens, but has little effect on others. More recently Boosalis et al. (1986) reviewed the literature on the effect of surface tillage on plant diseases. Surface tillage is somewhat different from no-tillage. Sprauge and Triplett (1986) defined surface tillage as follows: "Breaking, tearing, cutting, or otherwise loosening of the surface layers of soil with variously shaped disks or sweep or chisel cultivators prior to seed placement. Normally some plant residues are buried to shallow depths, while the remainder shield the soil surface as the crop becomes established."

Palti (1981) described and listed the effects of inoculum carried in stubble and crop debris on the infection of succeeding crops of wheat, barley, and maize. In all but a few cases there was an increase in disease in the succeeding crop. Apparently less research has been reported on the influence of tillage on nematodes, viruses and bacterial pathogens. Mizukami and Wakimoto (1969) noted that plowing under rice stubble between rice crops reduced sources of bacterial blight (*Xanthomonas campestris* pv. *oryzae*) inoculum. Sumner et al. (1981) concluded:

> Conservation tillage and minimum-tillage practices may increase, decrease, or have no effect on plant diseases. Leaving plant debris on the surface or partially buried in the soil may allow numerous pathogens to overwinter, or survive until the next crop is planted, but conditions favorable for biological control of plant pathogens may also be increased. Tillage practices directly influence physical and chemical properties of the soil, soil moisture and temperature, root growth, and nutrient uptake, and populations of vectors of plant pathogens. These factors in turn may influence the viability of plant pathogens and the susceptibility of the host.

Soil Solarization

Soil solarization is a soil disinfestation method using solar heating of soil. Weed control is often another benefit of the practice. As it has developed in Israel (Katan 1981, 1987) and California (Pullman et at. 1981), it uses thin, transparent, polyethylene tarps to cover the soil and trap solar energy. Palti (1981) noted that successive tillage operations can reduce the inoculum of some soilborne pathogens by solarization. Traditional farmers often plowed and harrowed land several times before planting, and may have been producing an effect similar to soil solarization that was effective against some soilborne pathogens. Such solarization effects may have been the basis for some traditional fallowing practices. Katan (1981) noted that the system is similar to other techniques for heating the soil by solar energy during summer seasons that have been practiced for centuries. Soil solarization may be appropriate for some developing countries.

Disposal of the plastic used in soil solarization may be problematic. In Costa Rica huge piles of used plastic are found near banana processing plants. The plastic contains insecticide and is used to cover the banana raceme during growth. Managers of such plants do not know what to do with the plastic and said that they are waiting for a solution to this serious environmental problem.

Traditional No-tillage Systems

Traditional farmers for centuries have been using agricultural systems similar to minimum or no-tillage systems. The systems did not include the use of herbicides; nevertheless weeds often were managed effectively and adequate yields were sustained by means of such practices as burning, long fallow periods, and mulching. For example, systems as diverse as the shifting or slash and burn system (Conklin 1961), the *tapado* system (Araya and Gonzalez 1987, Galindo et al. 1982), and the *marceño* or *popal* system of Mexico (Garcia-Espinosa 1987), and the planting of *quinua* and *tarwi* in the altiplano of Peru (Rengifo Vasquez 1987) have some characteristics of no-tillage systems. The use of pointed sticks (dibble sticks) by traditional farmers in Latin America for planting maize and beans is similar to the opening of a slit or punching a hole in the soil for planting in the no-tillage system. Rengifo Vasquez (1987) described the various systems for managing soils used in the Andes for centuries, many of which are still in use today. These systems are discussed in detail in Chapters 9 (Fallow), 10 (Fire), and 12 (Mulching).

The *marceño* or *popal* system is a most interesting tillage system. Large, swampy areas of the state of Tabasco, Mexico, were flooded for 6-8 months of the year. When these areas dried, the Chontales Indians cultivated maize. The dominant vegetation in these areas was popal grass (*Thalia geniculata*). In this highly productive system (*popal* system) a cultivar of maize (Marceño) was planted in holes (10-15 cm deep) in fields covered with popal grass that had been previously cut and allowed to dry. When the maize leaves began to emerge after planting, the 30-40 cm layer of dry grass was burned, but the maize seedlings survived and grew. Yields of 4-5 t/ha of maize were reported. The burning liberated nutrients and also eliminated weeds, insects and perhaps some plant pathogens. Soils in the system were under anaerobic conditions during flooding, which would eliminate many soilborne pathogens. Some evidence of biological control of soilborne pathogens was also found. Garcia Espinosa (1980b, 1985) inoculated maize soil from Tabasco that had a declining production due to soilborne pathogens and *popal* soil, which has a 30% organic matter content, and found evidence of suppression of *Pythium* spp. and *Rhizoctonia* spp. in the *popal* soils. Lumsden et al. (1981) found that *popal* soils were suppressive to *Pythium aphanidermatum, Sclerotinia rolfsii,* and *Rhizoctonia solani.*

Summary

Ancient peoples often had relatively sophisticated farming systems that used most of the tillage practices common today. Indigenous farmers in Africa, Asia, and Latin America are still using some of these practices in their traditional agriculture. Many of the traditional, most enduring systems of agriculture, -- for example the the slash and burn system, the slash/mulch system, the *tapado* system, and the *popal* system -- are essentially minimum or no-tillage systems. Recently farmers in the US have "rediscovered" the benefits of minimum tillage and its practice has increased dramatically in recent years. A major difference between traditional and modern minimum tillage is the use of chemical herbicides in the modern minimum tillage system.

A number of fungal pathogens survive in crop residues in minimum tillage systems; however, minimum tillage may increase, decrease, or have no effect on plant diseases. Survival of pathogen inoculum can increase disease, but conditions favorable for biological control may also be enhanced by minimum tillage. In addition to their many agronomic benefits, minimum tillage systems should be evaluated whenever feasible for their value in disease management. A combination of rotation with minimum tillage probably does the best job of managing plant diseases.

Investigators suggest that with improved conservation and no-tillage techniques a more intensive and profitable agriculture is possible in the developing countries of the tropics (Akobundu and Deutsch 1983, Greenland 1975, Lal 1975, 1981, Rockwood and Lal 1974, Warren 1983).

Soil solarization is a tillage technique that might be added to those practices appropriate for the farmers of some developing countries.

Crop Manipulations

18

Harvesting and Post-Harvest Storage

Many ancient civilizations (Arabic, Egyptian, Incan, Roman) had extensive systems of storage for agricultural products. Ancient authors made many recommendations concerning crop storage. The Greek Theophrastus (372-285 B.C.) had a section in his writings on proper storage of seeds (Orlob 1973). The Roman Varro (1934), who wrote in 116-27 B.C., stated that beans and other legumes keep well in an olive jar when sealed with a layer of ashes. The Greek treatise "Geoponica", compiled by Berytios (350-400 A.D.), noted that apples could be stored by wrapping them in seaweed or walnut leaves and putting them in a cool, dry place in an earthen container (Orlob 1973). Berytios also suggested that apples do not rot if stored among barley seeds. Storage practices in ancient India for various crops were discussed in Raychaudhuri (1964). It has been calculated that the storehouses of the Inca Empire contained a 3-7 years' supply of food consisting mainly of grain and freeze-dried tubers (National Research Council 1989b). Prehistoric man may have flamed seed as a protection against fungi and other pests, as suggested by Hopf (cited by Orlob 1973). Little is known about the actual magnitude of past crop losses due to microorganisms in ancient times.

The book "*Kitab al-Felahah*", written in the twelfth century by the Spanish Arab Ibn al-Awam (1988), has an entire section on the storage of various seeds and fruit. Some of his recommendations go back to even more ancient authorities. Ibn al-Awam wrote that in Southern Spain unthreshed wheat and millet stored better than threshed grain and he suggested mixing grain to be stored with dry leaves, chalk, or ashes. Storing seed for planting with ashes was also suggested for cereals and for vegetable seeds.

Many ancient peoples understood some of the important principles of grain storage. For example, in 1513 the Spaniard Alonso de Herrera (1988) wrote: "Pliny says that where there is no air insects can't reproduce, and I think that could be the reason, because they cannot live without air when they respire, and for this reason some make the storages so tight for wheat that in no manner can air enter." Alonso de Herrera added that wheat and millet could be kept for a decade in the best storages, and that this longevity was useful for fortresses and places that fear war. Columella (1988) described the construction and ventilation of the ideal granary and wrote that a granary should be as dry and cool as possible for grain preservation. Varro (1934) who lived 116-27 B.C., wrote: "wheat should be stored in granaries above ground, open to the draughts on the east and the north, and not exposed to damp air rising in the vicinity."

Storage Losses in Developing Countries

Fungi, insects, mites, and rodents all take their toll. In addition to reducing germination when crops are stored for seed, storage fungi can discolor seed, and cause heating and mustiness, which can make it useless for food. Storage fungi also produce mycotoxins toxic to humans and animals (Bean and Echandi 1989, Christensen and Kaufmann 1969, Mirocha and Christensen 1974, Purchase 1974). *Aspergillus flavus*, one of the most common storage fungi, produces dangerous aflatoxins (Williams and McDonald 1983) that are carcinogenic.

Many estimates have been made of postharvest losses in developing countries (Booth and Coursey 1972, Bourne 1977, Christensen 1979, Coursey 1983, Coursey and Booth 1972, Coursey and Proctor 1975, Hall 1969, National Research Council 1978, Proctor et al. 1981). The National Research Council (1978) stated: "Reliable studies indicate that postharvest losses of major food commodities in developing countries are enormous, in the range, conservatively, of tens of millions of tonnes per year and valued at billions of dollars." Christensen (1979) wrote: "It has been estimated that in some of the tropical countries as much as 30% of the harvested food grains is lost in storage." Hall (1969) estimated an average 20% post-harvest loss for grains, oilseeds, and pulses in developing countries. Losses of perishable foods such as vegetables and fruits are often higher, but little sound information is available. Coursey (1983) has estimated post-harvest losses in perishable crops to be between 10 to 30% in developing countries, according to commodity and specific conditions of storage. For example, as much as a third to a half of the bananas grown for fruit in tropical countries never reach a consumer. I have seen huge piles of rotting bananas, which did not meet standards for exports, in several banana-producing countries.

Storage Practices for Potatoes in the Andes

Potato storage in Peru probably dates back thousands of years. Traditional potato storage in Peru is described in detail by Rhoades et al. (1988) and Werge (1977). The Incas and even pre-Inca civilizations in Peru apparently had large, state-organized storage networks in addition to the stores of individual farmers (Cieza de Leon 1985, 1986, D'Altroy and Hastorf 1984, D'Altroy and Earle 1985, De la Vega 1966). According to calculations made by D'Altroy and Hastorf (1984), Inca storehouses in the Mantaro Valley of Peru had the capacity to store about 1,500,000 bushels of grain, tubers, or other food.

The ancient Indians of Peru developed another practice for long-term storage of potatoes. They freeze-dried bitter, frost-resistant potatoes (primarily *Solanum juzepczukii* and *S. curtilobum*) grown at 3,600 to 4,400 meters above sea level to produce *chuño* (Christiansen 1977, De Murúa 1987, Gade 1975, Mamami 1978, National Research Council (1989b), Werge 1979). After harvest, the potatoes were spread out to freeze at night when a heavy frost was expected. Men and women trampled the potatoes, and the water released by freezing was squeezed out. Black *chuño* was made by freezing the potatoes for several nights, trampling them during the day, and then drying them for five to ten days in the intense sunlight found at high elevations. White *chuño* underwent a similar process, but after being trampled the potatoes were placed in a stream of cold, running water. This process leached out dark colors and bitter glycoalkaloids and produced a more desirable product with a white color. Subsequently the white *chuño* was also spread out in a field and sun-dried. *Chuño* will keep for years in storage if properly dried. Other Andean tubers such as oca (*Oxalis tuberosa*) were also freeze-dried.

When the Spaniards arrived in Peru, Indians were producing *chuño* (Cieza de Leon 1984, De Acosta 1987, De la Vega 1966, Poma de Ayala 1987). Bernabé Cobo wrote in 1653: "There are some wild bitter potatoes called "*afora*" that are not eaten in fresh form, but that are good for *chuño*. The *chuño* is so hard it even keeps for years. It does not rot or deteriorate" (Christiansen 1977). According to Christiansen, in 1977 about 37,000 metric tons of *chuño* were produced in Peru by traditional farmers, so *chuño* is still important in the highlands. Christiansen also cites an ancient Inca proverb extolling the value of *chuño*: "Stew without *chuño* is like life without love."

Another Inca practice that reduced storage losses was to build storages at high altitudes in the Andes mountains, where low temperatures would prevent deterioration of stored products. Cobo (Mateos 1956) states: "The placing of these storages at high altitudes, was done by these Indians, in order that the contents of the storages were protected from water, humidity, and rotting." Extensive Inca storages were found near Cuzco and Huancayco in Peru. Both cities are at high elevations with cool temperatures. Research has found that the optimum temperature for storing potatoes is 4 °C. Low temperatures are also beneficial for grain storage and reduce insect and fungal attack (Christensen and Kaufmann 1969).

Although losses today in potato storages in the Andes are often high, most of the loss is perceived as due to insects. Rhoades et al. (1988), in a study conducted in a humid, warm valley in the Andes of Peru, cited storage losses of 47% for larger producers, 37% for medium sized producers, and 16% for small producers. Such high losses would probably not occur in the cooler climates more commonly used for potato production. Losses were due to insects, dehydration, rotting, and rodents. It is noteworthy that the small producers, probably the most traditional, had the smallest losses. What caused the potatoes to rot in these studies unfortunately was not discussed, but fungi and bacteria were undoubtedly of importance. Rhoades et al. also noted that farmers prefer cultivars with long dormancy periods for storage. Today, traditional farmers in the Andes of Peru sometimes prevent insect damage by storing potatoes with leaves of muña (*Mintostachus mellis*) or Eucalyptus trees. Both plants are aromatic and have insect repellent qualities. Insect damage often provides access for fungal infection.

Peasant farmers in the Andes (personal communication -- Gordon Prain) have several additional methods of reducing storage losses. For example, rather than being discarded, poor quality or diseased potatoes are often fed to animals. Also, potatoes with certain types of rot are cooked in a temporarily constructed earth oven called a "*pachamanca*", and the special taste of such potatoes is highly appreciated. There are processed potatoes called *tocosh* in Peru that are produced by allowing potatoes to ferment for up to six months in a stream. After removal from the stream the *tocosh* potatoes are boiled as needed.

During the 1950s, when I lived in Colombia, one of the most popular potato varieties was Tuqerreña, which had a dark purple skin. Such purple cultivars, which often have a long dormancy period, did not turn green when exposed to light for long periods. This may have been an important characteristic for rustic potato storages, as green potatoes are difficult to sell and produce solanine, a highly poisonous glycoalkaloid, which also imparts a bitter taste to potatoes. However, farmers in Peru shunned the dark-skinned variety Mariva because it produced the bitter tasting solanine when exposed to light, even though it did not turn green (personal communication -- Gordon Prain).

Traditional Storage Practices in the Tropics

Traditional farmers over the centuries developed a wide variety of practices for storage of agricultural products, and our modern storage techniques make use of much of this heritage. A National Research Council publication (1978) gave information on indigenous storage of a number of major crops. For example, traditional millet storages in Mali and Senegal were described, in which losses were said to be lower than in more modern large central warehouses. One of the traditional storages described was built of rock, wood, and clay mixed with millet chaff and had a conical thatch roof. The foundations were elevated on a wooden structure that kept the storage off the ground, thus protecting the millet from moisture. Some millet was stored on the head and some in bulk with ash. The ash was scattered on the floor of the storage, rubbed into the walls, and mixed with the threshed grain. Fungal damage in these stores was reported to be negligible.

Jurion and Henry (1969) described storage practices of indigenous farmers in Zaire and suggested that storage of cassava, sweet potatoes, and bananas in Zaire was not a major problem there, as these crops were usually consumed as needed. Consumption "as needed" probably is common, but considerable food is lost in the tropics when more is produced than can be consumed. Cassava and bananas are dried in the sun and converted to chips or flour if needed for sale or barter. Jurion and Henry added that in the humid climates of tropical West Africa grain storage was difficult. Nevertheless, traditional African farmers have shown considerable ingenuity in storing their grain and protecting it from fungi and insects. Jurion and Henry wrote:

> Where the climate is humid, grain that is to be eaten is kept inside the huts, protected by the heat and smoke in such a way that air can circulate through it, while maize -- if it has not been eaten green -- is stored on a rack hanging 1 or 1 and 1/2 m above the hearth. These measures do not provide complete protection and extra care has to be taken with seed grain; this is first dried and then put into calabashes or earthenware pots which are carefully sealed and stored in rows over the hearth.

Farmers in West Africa stored seeds of onions, tomatoes, okra, maize and beans above fires in their houses, and in drier regions ash was mixed with shelled beans and Bambara groundnuts before storage (Zehrer 1986). Wood ash and dried citrus leaves were mixed with pulses in Sri Lanka (Upawansa 1989). Miracle (1967) listed four principle methods for crop storage in the Congo Basin: in the farmer's hut, in granaries, in hermetically sealed pots, and in packages suspended in trees. He wrote:

> When foodstuffs are stored in the farmer's hut, they are usually put on a shelf placed on supports over the hearth, or suspended from the roof. This keeps their moisture content low, and the regular applications of smoke reduce weevil damage. Evidence available on maize strongly suggest that storage losses incurred with grain stored in this fashion are less than 5% in the Congo Basin.

Traditional granaries were described as cylindrical or rectangular and raised off the ground to allow for air circulation. Fires were occasionally started under the granaries to further dry the grain. Miracle (1967) mentioned that two groups (the Azandi and people of the Ubangi district of Zaire) used the space under granaries as kitchens. The heat from cooking fires dried the grain. According to Conrad Bonsi (personal communication) this practice was also common in Ghana (Figure 18.1). Miracle described numerous other indigenous storage practices used in Zaire.

De Schlippe (1956) described the storage and processing practices of the Zande people of Central Africa, many of which are similar to those described above. Sagnia (1989) reported that millet heads are hung over kitchen fires in the African Sahel, and that storing millet in the head, as opposed to storing it as threshed grain, was better, as the glumes acted as protective devices against pests. Threshed seeds were mixed with sand and wood ash for storage.

Storage Practices for Maize

The traditional storage practices for maize in both North and South America were described by Weatherwax (1951). He related that about 1621 Captain John Smith of the Virginia colony discovered what the local Indians already knew, i.e. that corn dried in the husk and subsequently stored in the husk has much less insect and fungus damage than shelled corn. Weatherwax described several methods of traditional maize storage by American Indians, such as storage in caves, in baskets covered with sand, in bags, pottery vessels, and bark boxes, in cribs made with poles (Iroquois), and in the husk festooning their permanent houses on rafters, beams, and walls. The maize storage methods of several American Indian groups were also described by Barrerio (1989), while Hurt (1987), Parker (1910), and Lewandowski (1989) described maize storage in underground pits by the Iroquois Indians. According to Buffalo Bird Woman (Wilson 1987), Hidatsa Indians also stored maize and squash seeds in underground cache pits lined with a kind of slough grass that would not become moldy. The openings were concealed to hide the cache from enemy Indians such as the Sioux. A rather unique storage practice was followed by the Hidatsa Indians of North Dakota in that they gathered, dried, and stored maize smut (*Ustilago zeae*), which they called *mapedi*, for future consumption (Wilson 1987). Maize smut is commonly eaten in Mexico, and today canned corn smut is available there in commercial stores.

Poma de Ayala (1987), a Peruvian Indian writing in the sixteenth century, described maize storage during the time of the Inca empire. After harvest, the highest quality seed was saved for planting, a somewhat lower quality was saved for consumption, and the lowest quality seed was used for making *chicha* (a fermented drink). After classification, maize was stored or consumed according to its quality. Cobo (Mateos 1956), a Spanish priest writing in the seventeenth century, wrote that on the dry coast of Peru, maize was often buried in the sand for storage. He also noted that there, where it almost never rains, wheat left in bundles without threshing lasted longer than wheat in storage.

Describing the storage of maize by traditional farmers in the Choco of Colombia, West (1957) mentions that maize is hung in rafters near the hearth in round bundles (*ensartos*), where smoke will thoroughly dry it. This is also a common practice among traditional groups in Africa (Miracle 1967). Upawansa (1989) describes the storage of maize and millet on the rack of the fireplace in Sri Lanka.

Doblando la Mazorca (Twisting Down the Ear)

Drying maize is an especially difficult problem in the warm, humid tropics. After the maize grains have filled out, it is common practice for traditional farmers in much of Latin America to bend the stalk or the ear so the tip of the ear hangs down (Figure 18.2). This twisting down of the stalk or the ear is called "*doblando*" or "*doblando la mazorca*" in Spanish. Farmers found that by using this practice the grain is protected from rain, it dries better on the plant in the sun than in storage, is less likely to be blown down by the wind, is less accessible to rats and birds, and reaches such a low moisture content that storage deterioration is greatly reduced. Montoya and Schieber (1970) sampled maize in Guatemala that had been bent down, and found only 1.0% of the grain was damaged by fungi, compared to a mean of 14.5% of the grain damaged from similar maize plants whose ears had not been bent down.

Doblando appears to be an ancient practice still common today. Mayan frescos at Tulum, Mexico, depict a doubled or bent maize stalk (Turner 1978a). Patiño (1965) cites several references to *doblando,* stating that it is a common practice in much of Latin America. The Museo Nacional de Culturas Populares of Mexico (1982) reports its continued use by Mexican traditional farmers. Stadelman (1940) and Carter (1969) report its use in Guatemala. Emerson (1953) and Morley and Brainerd (1946) wrote that *doblando* is a common practice of the Mayan Indians. The Mayan Zinacantecos in highland Chiapas use the practice (Cancian 1972). I have seen the practice in several countries of Central and South America and also in Spain. Weatherwax (1954) found references describing its use by Mexican Aztecs in the sixteenth century. He cites Friar De Sahagun (who went to Mexico in 1529) as listing the following work for an Aztec maize farmer:

> The duties of the farmer are: to fill up the holes where maize is planted, to heap the earth around the young plants, to eradicate the grass, to thin out the plants and remove the small ears and ear suckers and tillers so that the plants will grow well, to take off the green ears at the proper time, *to break over the stalks at maturity* and harvest the corn when it is dry, to husk the ears and knot the husks together or fasten the ears together in strings, to carry the harvest home and store it, to break up the stalks which have no ears and to shell the grain and clean it in the wind.

Storage Practices for Roots and Tubers

Cassava is perhaps one of the most perishable tropical crops, as a physiological deterioration discolors the roots a few days after harvest, making them unacceptable for consumption. One of the most important practices used by traditional farmers for storing cassava is to leave it in the ground, and harvest it only as needed. Cassava will last for two or three years in the ground, but with increasing age, quality is reduced. Many different dried cassava products are prepared in various areas of the tropics. Chips and various kinds of flour or meal are prepared from dried cassava. After peeling and grating roots, a dried or fried meal called "*gari*" is prepared from fermented pulp in West Africa. *Farinha da mandioca* is prepared in Brazil in a similar manner, but is usually not fermented. These products will last for several months if thoroughly dried; however, fungi may attack them if moisture levels exceed 12-13% (National Research Council 1978).

The processing of plantains, cassava, yams, and sweet potatoes by indigenous peoples involves many complex techniques, some of which have been described by Miracle (1967). Bitter cassava with high contents of HCN must be processed to remove the toxic cyanide. In parts of tropical Africa this is done by soaking the roots in running water or specially made mud holes. Miracle found the soaking period to vary from six hours to 2 & 1/2 weeks. Some groups peel cassava before soaking and some do not. After soaking, the roots may be boiled at once, or immediately pounded into flour and boiled. Several cassava products and methods of processing found in Zaire are described by Miracle. He also describes in detail the preparation of various alcoholic beverages: "Wines are made from palm and bamboo sap, sugar cane, bananas, plantain, and pineapples. Beers are made from maize, millet, sorghum, and manioc." The preparation of various fermented beverages, such as beer, wine, or *chicha,* is a short-to long-term storage method widely used by traditional farmers.

Yams are stored in West Africa in "yam barns" (National Research Council 1978). Generally these consist of a wooden framework about two meters high to which the yam tubers are tied (Figure 18.3). The poles supporting the frame work are often of wood that will take root and grow when placed in the ground. This prevents termites from attacking the poles and also from attacking the yams. The poles may even provide some shade after sprouting, but usually the yam barn has a thatched roof. A serious storage problem of yams is caused by nematodes that may also increase the susceptibility of the yams to storage fungi (Ekundayo and Naqvi 1972). Several fungi cause storage rots of yams (Ekundayo and Naqvi 1972, Ogundana et al. 1971). Yen (1974a) described the storage practices of the Kakoli people of New Guinea, who store about 25 pounds of sweet potato roots in holes in the ground sealed with a turf plug.

Storage Practices for Rice

Much of the world's rice is grown in warm, humid areas, and so, unless rice is dried to a low enough moisture content, it can be damaged by fungi and insects in storage. Traditional growers often store rice in bundles of panicles. Rice is usually stored with the husk still on the grain (paddy). The National Research Council (1978) stated: "The protection afforded the kernel by the husk against insect, fungal, and even rodent attack, as well as the problem of storing poorly milled rice, accounts for the fact that the bulk of harvested dried paddy is stored in this form before milling." Barnett (1969) described the rice storage practices of the Ibaloi in the Philippines. The rice is stored in bundles in the rafters of their houses. During the rainy season rice is kept above the cooking hearth, becoming black with soot, and is not threshed until needed for consumption. Conklin (1980) described and illustrated the practices of the Ifugao in Northern Luzon who store rice bundles in special storage structures.

Summary

In a world where millions suffer from hunger and starvation, it is shocking to contemplate the enormous losses that occur after crops are harvested. Many of the estimates of losses are so high that they are difficult to believe. For example, perhaps 30% of the harvested food grains in some tropical countries are lost in storage (Christensen 1979), and some have estimated that in some cases as much as 50% of the fruits and vegetables produced in the tropics are never consumed. My thirty-seven years of experience, travel, and observations in the tropics tend to confirm the above figures. Although considerable effort and funding has gone into devising methodologies and computer technology for measuring losses with considerable precision in developing countries, the actual measurement of storage losses in the tropics is neglected. Little reliable information on actual losses can be found. In addition to the loss of foodstuff caused by fungi, insects, and rodents, another concern is that storage fungi produce mycotoxins, which are toxic or carcinogenic to humans and animals. The importance of mycotoxins in foods in the tropics can only be conjectured, but mycotoxin toxicology is undoubtedly serious.

Traditional farmers have developed a variety of ingenious and effective practices for crop storage over the centuries. Freeze-drying potatoes in the Andes, bending down the stalks or ears of maize, fermenting crops in various ways, storing grain over cooking fires, and using hermetically sealed containers for grain storage are but a few examples of the postharvest practices of traditional farmers. These often rustic practices need careful study and evaluation as their capital and technological skill requirements are generally low, and adoption often requires little restructuring of traditional societies.

19

Multiple Cropping

There is ample historical evidence that multiple cropping, a common practice in traditional agriculture systems in most tropical areas today, is an ancient practice. It may have been used long before the time of Christ in Egypt and Babylon and was used about 1,000 B.C. in India (Dalrymple 1971). Watson (1983) suggested that multiple cropping systems reached the Near East region from India, perhaps even before the rise of Islam. As early as 1918, Butler wrote that intercropping in India helped to control plant diseases. The Romans' use of multiple cropping was described by White (1970). During the Ming Dynasty (1644-1840) multiple cropping extended all over China (Youtai 1987). Various forms of multiple cropping were probably used by ancient traditional farmers long before monoculture become common.

Associations Used in Multiple Cropping

Innumerable crop combinations are used in the tropics. The trilogy of maize, beans, and squash was, and still is in many areas, commonly used by traditional farmers throughout North and South America. Estrella (1986) reported that maize was planted with beans in mounds (*caballónes*) in Ecuador in 1573, and describes extensive use of multiple cropping in Ecuador. The number of different crop associations planted by traditional farmers is often astonishing. Kirkby et al. (1980) found 100 different crop associations grown in a 30,000 ha traditional project area near Riobamba, Ecuador (Figure 19.1). I have seen maize, common beans, faba beans, tarwi, quinoa, and squash all interplanted in the same field in this region of Ecuador.

Importance and Prevalence of Multiple Cropping

Much of the traditional agriculture in developing countries involves multiple cropping strategies. While traveling in the tropics, I have seen dozens of different crop combinations. Jodha (1979) wrote that 50-80% of the crops in rainfed areas of the developing countries were intercropped. Leihner (1983) related that 40 and 50%, respectively, of all cassava grown in the Americas and Africa was intercropped. Okigbo and Greenland (1976) found that 27% of cassava, 76% of maize, 90% of millet, 96% of peanut, and 99% of cowpea were produced in multiple cropping systems in Nigeria. In Uganda 85% of maize, 50% of cassava, 81% of beans, and 76% of pigeon peas were intercropped, according to Jameson (1970). Francis (1988) reported that 90% of the beans in Colombia, 80% of the beans in Brazil, and 60% of all maize in the American tropics were in multiple cropping systems. In the Brazilian Amazon Kass (1978) reported 99% of the maize and rice and 67% of the cassava were grown in mixed stands. Kass gave additional figures on the frequency of intercropping in different countries in his review of intercropping. Crop species in slash and burn agriculture are often intercropped. Salik (1989) found that the Amuesha Indians of the Peruvian Amazon grew 96 different crop plants, and that intercropping was found in all of their cropping systems. Jurion and Henry (1969) and Miracle (1967) listed many of the complex crop associations found in Zaire. Aiyer (1949) and the International Center for Research in the International Crops Research Institute for the Semi-Arid Tropics (ICRISAT 1981) gave detailed descriptions of multiple cropping in India. Jodha (1949) studied intercropping in six villages in the semi-arid areas of India and found that from 18% to 83% of the farmed area was intercropped. Tropical crops such as coconuts, rubber, and oil palm were usually widely spaced and intercropped with other crops, which were planted beneath (Nair et al. 1974, Nair 1983, Nelliat and Ji 1976, Pinchinat et al. 1976). The above figures provide striking evidence on the prevalence and importance of intercropping in developing countries, especially those in the humid tropics.

Dalrymple (1971) wrote that the Chinese probably have more land in multiple cropping than the rest of the developing countries combined. Harwood and Plucknett (1981) mentioned that they saw literally hundreds of different crop combinations while traveling in China. They defined intercropping as: "the growing of more than one crop in the same field at the same time." Dalrymple (1971) and FAO (1980) describe the innumerable combinations of vegetables made by traditional Chinese farmers.

Effects on Yield

Many studies have shown that total production is generally higher in intercropping (Evans 1960, Kass 1978, Liebman 1987, Mutsaers 1978, Trenbath 1974, Trenbath 1986). The Land Equivalency Ratio (LER) is commonly used for the evaluation of the productivity of mixed cropping. Harwood (cited by Beets 1982) defined the LER as "the total land required using monocultures to give total production of the same crops equal to that of one hectare of mixed crop." Kass (1978) and Beets (1982) reviewed the literature on the effect of polycropping on yields.

Labor Requirements

Data reported by Gomez and Gomez (1983) from Taiwan and Indonesia show clearly that multiple cropping increases as farm size decreases. They also cited data from Asia showing that labor requirements increased as the number of crops increases. They suggested that as many countries are trying to find employment for their rapidly increasing populations, multiple cropping is an excellent strategy for intensifying land use and absorbing excess farm labor. Jodha (1979) reached similar conclusions and added that intercropping assures greater and more even distribution of labor throughout the year.

Research on Multiple Cropping

When the cropping systems program of CATIE (Centro Agronomico Tropical de Investigacion y Enseñanza), Turrialba, Costa Rica, began looking at the innumerable combinations and permutations of crops possible in intercropping experiments, they called their work "el experimento de los locos" (the experiment of the crazy people), I was told, because of the complicated spatial and sequential arrangements possible and the confusing statistical and agronomical problems encountered. For example, they described a comprehensive study in which one field experiment in 1973 in Turrialba had 54 crop patterns tested at four levels of technology, thus forming 216 different "cropping systems" (Pinchinat et al. 1976). Research on multiple cropping is complicated, difficult, expensive, and time consuming. Nevertheless, considering the importance of multiple cropping in developing countries, much more research is needed to improve utilization of the practice.

Literature on Multiple Cropping

No attempt will be made in this paper to seriously review the extensive literature on multiple cropping as many books (Beets 1982, Francis 1986, Gomez and Gomez 1983, ICRISAT 1981, Papendick et al. 1976, Steiner 1984), articles and bibliographies (Altieri and Liebman 1986, Dalrymple 1971, Dover and Talbot 1987, Grigg 1974, Harwood 1979, ICRISAT 1981, Kass 1978, Liebman 1987, Litsinger and Moody 1976, Monyo et al. 1976, Okigbo and Greenland 1976, Pinchinat 1976, Polthanee and Marten 1986, Ruthenberg 1980, Soria et al. 1975, Sumner et al. 1981, Tarr 1972) are available on the subject. Allen (1989) and Altieri and Liebman (1986) discussed the effects of multiple cropping on plant diseases.

Definitions of Multiple Cropping

Polycropping or multiple cropping terminology is extensive and rather confusing. In this chapter the terms intercropping and polycropping will be used interchangeably, according to the definition of Vandermeer (1989) for intercrop (polyculture), unless the mixed system is clearly different: "the cultivation of two or more species in such a way that they interact agronomically (biologically). Intercrops can be of four flavors - mixed, row, strip, or relay." Many types of multiple cropping are discernible in traditional agriculture, and different authors have given the terms listed below specific meanings (Francis 1986, Harwood 1979, Kass 1978, Leihner 1983, Vandermeer 1989). The terms "crop mixtures", "crop combinations", and "crop associations" are also commonly used. Although some of the terms below are self-explanatory, unfortunately some are not used consistently or correctly in the literature. Authors often do not clearly describe the nature of the intercropping. Clarification of this confusing terminology is beyond the scope of this book, and thus the reader is referred to the above references, especially Francis (1986).

1. double cropping
2. intercropping
3. interculture
4. interplanting
5. maximum cropping
6. mixed cropping
7. multiple cropping
8. polycropping

9. relay intercropping
10. relay planting
11. row intercropping
12. sequential cropping
13. strip intercropping
14. triple cropping

Advantages of Multiple Cropping

Francis (1988) discussed the efficiencies of resource use by multiple species and noted that several interactions may occur among or between crop species: 1) mutual inhibition; 2) mutual cooperation, and 3) compensation. For multiple cropping to be advantageous to the traditional farmer there must be some complementarity between species (Francis 1988). This may be an increased efficiency in the use of light, water, or nutrients, especially nitrogen. There also may be better management of weeds, insects, or plant pathogens.

Wilken (1987) summarized the value of multiple cropping as regards space as follows:

> By incorporating crops that occupy different horizontal and vertical spaces, multiple cropping makes efficient use of field space. When multiple cropping is coupled with scheduling, the use of spaces becomes more continuous. A further step is to set aside separate areas where crops pass germination and early seedling stages under special care and where they occupy only a fraction of valuable field space. Increased productivity comes at the price of increased labor, however. The highest crop densities are found in farming systems that use the least fossil fuel or even supplements of animal energy.

Wilken also noted that intercropping reduces erosion because of the complete plant cover, that tall plants may provide support for vines, that leguminous crops provide nitrogen, and that intercropped fields usually have fewer pests and pathogens on aerial plant parts than plants in monocultures. Gliessman (1986) discussed the various plant interactions that can occur in multiple cropping systems. Radiation and microclimate relationships in multiple cropping systems were discussed by Allen et al. (1976) .

Although the disadvantages of multiple cropping have not been extensively discussed, several negative effects may occur. Multiple cropping often requires more labor, allelopathic effects may occur, herbicides can be used on intercrops only with great care, shading effects may reduce yields, and there is sometimes increased occurrence of diseases and other pests. The advantages and disadvantages of multiple cropping have been thoroughly reviewed by Kass (1978) and Francis (1986).

Flooding and Multiple Cropping

Flooding may play an important role in some multiple cropping systems. A common cropping system in Taiwan is two crops of rice followed by two to three upland crops of vegetables, with 4-5 crops per year (Anon 1974). The anaerobic or near anaerobic conditions produced by flooding are known to reduce populations of many soilborne pests and pathogens (Cook and Baker 1983, Palti 1981, Stolzy and Sojka 1984, Stover 1979), and thus may have important positive effects. The effects of flooding on diseases are more completely discussed in Chapter 11.

Household Gardens and Multiple Cropping

Some of the most extreme examples of diversity in multiple cropping systems are found in the household gardens common all over the tropics. The household gardens in West Java described by Michon et al. (1983) are one example. These small gardens may constitute from 15 to 50% of a village's land available for cultivation, and plants for food, timber, firewood, medicine, and ornamentals are grown. A great diversity of species is grown in the gardens, with some villages using up to 250 crop species. This diversity has important implications for the severity of diseases in the gardens, as pests and pathogens are less able to build up to destructive proportions on the few isolated plants of each species. Household gardens are more thoroughly discussed in Chapter 20.

Increased Disease in Multiple Cropping

The literature has numerous references to increased disease in intercropped plantings. More soybean rust was reported when soybeans were interplanted with maize at IRRI (1973). More rust (*Uromyces appendiculatus*) and Cercospora leaf spot (*Cercospora canescens*) were reported on mungbeans intercropped with maize by researchers at IRRI (1973). Experiments at IRRI (1979) showed that closer spacing of maize in intercropping experiments with mung bean increased powdery mildew (*Erysiphe polygoni*). Cercospora leafspot of peanuts (*Cercospora canescens*) was more severe when peanuts were grown with maize (IITA 1975). When rice was intercropped with maize or cassava more brown spot (*Cochliobolus miyabeanus*) was found than when rice was grown alone Kass (1978). Van Rheenen et al. (1981) noted more white mold (*Sclerotinia sclerotiorum*) when beans were intercropped with maize. National Academy Sciences (1968) described increased incidence of *Verticillium* spp. when olives, almonds, apricots, or avocados were intercropped with tomatoes. For example, *Verticillium* incidence in established olive plantings is normally low, but when olives are interplanted with tomatoes, the amount of *Verticillium* inoculum in the soil increases rapidly, and thus *Verticillium* incidence in olives increases. Blanco-Lopez et al. (1984) reported similar results when olives were intercropped with cotton or vegetables. Some of these disease increases may be due primarily to shading by the taller maize plants and a more humid microclimate in the vicinity of the shorter crops. Crop debris from previous intercrops may serve to overseason inoculum, and this may also increase the severity of some diseases.

An unusual case of increased disease in intercropped maize and cowpeas was related by Allen and Skipp (1982). Maize pollen, which frequently dusts the leaves of the plants with which it is intercropped, stimulated the germination of the conidia of *Colletotrichum lindemuthianum* (anthracnose of cowpea) thus causing increased disease.

Decreased Disease in Multiple Cropping

There are far more references in the literature reporting decreased disease in mixed cropping associations. Table 19.1 provides references.

Allen (1989) analyzed some of the possible reasons for both an increase and a decrease in diseases in intercropping. Some of the factors he lists as responsible for decreased disease are:

1. Physical barriers to the dissemination of aerial pathogens or their vectors
2. Altered microclimate: attributable to shading
3. Altered microclimate, altered relative humidity
4. Host-pathogen interactions: induced resistance
5. Spacing: increased distance between plants
6. Pollen grains affecting spore germination

TABLE 19.1 . References Citing Decreased Diseases in Intercropping Situations

Author(s)	*Crop Combination*	*Disease(s)*	*Pathogen*
Autrique and Potts 1987	maize/potatoes	bacterial wilt of potatoes	Pseudomonas solanacearum
Autrique and Potts 1987	beans/potatoes	bacterial wilt of potatoes	Pseudomonas solanacearum
Ene 1977	cassava/ maize/melons	cassava bacterial blight	Xanthomonas campestris pv. manihotis
Farrell 1976	peanut/bean	rosette virus	virus
Harwood 1979	maize/ soybean/rice	maize downy mildew	Peronosclerospora maydis
Keswani and Mreta 1982	sorghum/ mungbean	powdery mildew	Erysiphe polygoni
Larios y Moreno 1977	maize/cowpea	ascochyta blight of cowpea	Ascochyta phaseolorum
Larios y Moreno 1977	cassava/maize	superelongation of cassava	Sphaceloma manihoticola
Larios y Moreno 1977	cassava/maize	cassava dieback	Glomerella cingulata
Moreno y Mora 1984	maize/bean	rust of bean	Uromyces appendiculatus
Moreno 1975	maize/cowpea	ascochyta blight of cowpea	Ascochyta phaseolorum
Moreno 1979	maize/cowpea	beetle-transmitted virus of cowpea	virus

(Continues)

TABLE 19.1 (*Continued*)

Author(s)	*Crop Combination*	*Disease(s)*	*Pathogen*
Moreno 1979	maize/bean	rust of bean	Uromyces appendiculatus
Moreno 1979	maize/ cassava/beans	angular leaf spot of beans	Phaeoisariopsis griseola
Moreno 1979	maize/potato/ beans	angular leaf spot of beans	Phaeoisariopsis griseola
Moreno 1979	cassava/bean	angular leaf of bean	Phaeoisariopsis griseola
Moreno 1979	sweet potato/ bean	angular leaf spot of bean	Phaeoisariopsis griseola
Mukiibi 1982	peanut/bean	rosette virus	virus
Soria et al. 1975	cassava/beans	rust of beans	Uromyces appendiculatus
van Rheenen et al. 1981	maize/bean	anthracnose,	Various pathogens: ascochyta blight, scab, halo blight, common blight, scab, black node diseases, and common mosaic of bean

Sumner et al. (1981) listed air movement, temperature, humidity, and light as factors that might be affected in the microenvironments produced by intercropping. All of these factors may affect disease incidence and development. Ene (1977) reported that intercropping significantly reduced the severity of cassava bacterial blight (*Xanthomonas campestris* pv. *manihotis*) in Nigeria. Mulches also reduced the severity of cassava bacterial blight by reducing soil splashing. In Colombia J. Carlos Lozano (personal communication) found that when cassava was intercropped with maize, cassava was protected from cassava bacterial blight primarily by reduction of wind-splashed rain.

The work of Autrique and Potts (1987) showed that intercropping potatoes with maize and beans reduced the incidence and rate of potato bacterial wilt (*Pseudomonas solanacearum*) development in potatoes. The reduction was affected by increasing the distance between individual potato plants and by the presence between potatoes of roots of other crop species. They concluded: "the use of low plant densities and crop association, as presently practiced by many farmers in developing countries, is an efficient and complementary means of aiding in the control of the disease."

Harwood (1979) wrote that traditional farmers in Southeast Asia grew maize in rows 2-3 m apart and intercropped it with other crops such as mung beans, rice, peanuts, or soybeans. Intercropped maize had little downy mildew (*Peronosclerospora maydis*), but when maize was grown as a monocrop in dense stands downy mildew was often highly destructive. Sastrawinata (1976) also noted that the density of planting maize significantly affected downy mildew levels provided the infestation was not extreme. The space between the maize plants allowed air movement that caused the maize plants to dry off before infection by *P. maydis* could take place.

The association of maize and beans is one of the most common crop associations in the tropics, especially in Latin America. In Ecuador as early as 1573, Estrella (1986) reported that two maize seeds and one bean seed were planted in the same hole in mounds (*caballónes*). Moreno and Mora (1984) found the incidence and severity of bean rust was less when beans were grown in association with maize than when beans were grown in a monoculture. Van Rheenen et al. (1981), working in Kenya, compared bean monocultures with beans grown in association with maize. They found that beans grown with maize usually had fewer diseases and insect pests. Bean diseases reduced in this association were halo blight (*Pseudomonas syringae* pv. *phaseolicola*), bean common mosaic, anthracnose (*Colletotrichum lindemuthianum*), common blight (*Xanthomonas campestris* pv. *phaseoli*), scab (*Elsinoe phaseoli*), mildew (*Erysiphe polygoni*), and Phoma black node disease. Van Rheenen et al. suggested that the associated maize crop caused temperatures to decrease and humidity to increase, that it reduced light, and that the maize formed an umbrella over the bean plants and thus probably maize formed an umbrella over the bean plants and thus probably prevented spreading of spores by splashing. The maize also acted as a wind break, which reduced the spread of spores by wind. However, one disease, white mold of beans (*Sclerotinia sclerotiorum*), was more severe in the maize/bean intercrop.

Summary

With so many different types of multiple cropping found in traditional farming systems, it is difficult to generalize with any degree of accuracy about disease management through the use of multiple cropping. Whether diseases increase or decrease in crop associations depends on numerous factors affecting the crop, the pathogen, the macroenvironment, and the microenvironment. Recommendations and generalizations on multiple cropping should be thoroughly tested before dissemination to farmers. Often site-specific recommendations will be necessary. Time-tested local practices should serve as the first guide to recommendations. Nevertheless, most of the literature indicates that there is less disease in most types of crop associations than in monoculture.

20

Multistory Cropping and Traditional Household Gardens

Perhaps the most important architectural manipulation used in traditional farming systems is the stratified arrangement of crops commonly known as "multistory cropping." The vertical (height) distribution of species within such systems constitutes the architectural structure. Useful information on multistory cropping in traditional agricultural systems is found in Beets 1982, Christanty et al. 1986, Fernandes et al. 1984, Geertz 1963, Gliessman 1988, Harris 1971, Michon et al. 1983, 1986, Wiersum 1983.

There are innumerable permutations and combinations of crops used in the tropics, depending on climate, altitude, soil, and local traditions. Tropical crops such as coconuts, rubber, and oil palm are usually widely spaced, and thus there is often room for other crops to grow beneath them (Nair et al. 1974, Nelliat and Ji 1976, Pinchinat et al. 1976). Beets (1982) discussed multistory cropping with permanent and annual crops combined and multistory cropping with permanent crops only. He cited several examples of the various crop combinations used in such systems. Combinations of cacao and rubber and rubber and coffee are examples of multistory cropping with permanent crops only.

Household Gardens

Gardens variously called backyard gardens, home gardens, household gardens, kitchen gardens, village gardens, and dooryard gardens (Christany et al. 1986, Gliessman 1988, Gonzalez 1985, Kimber 1973, Michon et al. 1983, Niñez 1984, 1985, Sauer 1969, Wilken 1971, Wiseman 1978, Yen 1974b) are found around the homes of many traditional farmers in the tropics. Anyone who has traveled outside of the large cities in the humid tropics has seen the omnipresent backyard gardens. The gardens are usually multistoried systems dominated by tall trees and forming layers of vegetation, and in some tropical locations they closely mimic the natural forest

Indonesian Household Gardens

Abdoellah and Marten (1986), Christanty et al.(1986), Michon et al. (1983), Palte (1990), Terra (1958), and Wiersum (1983) described the several types of gardens found in Indonesia. Wiersum listed three types:

1. Home gardens -- fenced areas around individual homes in which various tree species and annual and perennial crops are cultivated together. Small animals are often included.
2. Mixed gardens -- these are usually outside the village, and are dominated by planted perennial crops, primarily trees; however, annual crops are cultivated beneath the trees.
3. Forest gardens -- found on private lands outside the villages. They consist of trees, which are either spontaneous or planted, and sometimes additional perennial crops occur.

Christanty et al. (1986) described in detail home gardens in Indonesia and stated that about 20% of the 4.4 million hectares in West Java were used for home gardens and 16% for a mixed garden/tree planting system. The average household farm size was only 0.5 hectares, and the population density was 700 inhabitants per square kilometer. The home gardens were highly diverse and contained many different species. The authors cited a study in which Hadikusumah as stated that he found 112 species in kebun-talun (a rotation system between a mixed garden and a tree planting) and 127 species in home gardens. The plants found included medicinal, ornamental, spice, vegetable, fruit, firewood, building material, and various cash crops. As Christanty et al. had observed: "The high diversity of plants in home gardens and kebun-talun form a complex horizontal structure, while a mixture of annuals and perennials of different heights forms a vertical structure."

Michon et al. (1983) discussed the village gardens in West Java, which were first described in the tenth century. Small in size (often less than 0.1 hectare), the gardens were nevertheless important in feeding the dense populations of Java. Such gardens constituted from 15 to 50% of the land available for cultivation for each village. Over 70 plant species were grown in the gardens, including plants for food, timber, firewood, medicine, and ornament. A striking diversity of species was used (some villages were reported to use up to 250 crop species), and this has important implications for the importance and severity of diseases in the gardens. Pesticides were seldom used or needed. Various animals were also important constituents of the gardens; they grazed, or were fenced within the garden and fed with garden products. Their waste contributed to nutrient cycling in the gardens. Fish were also found in ponds in some gardens; they were fed vegetable and animal waste.

Michon et al. (1983) stated that in these small gardens each plant received individual care. The gardens imitated the tropical forest ecosystems of Java. Each plant had its "place" in the garden, and the physical arrangement (horizontally and vertically) was sophisticated, taking advantage of the available solar energy and the tolerance of individual species to shade. The upper layer or top crop canopy utilized sunlight tolerant species; and as the gradient of light and humidity changed vertically, different species were grown in their proper "niches." The authors reported that traditional gardeners had reliable ecological knowledge, that allowed them to fit plants into sites fitting their various requirements. Abdoellah and Marten (1986) found 235 useful species of plants grown in household gardens in a survey in West Java. The average garden had 20 species while the average upland field had eleven species.

Household Gardens in the Americas

Wilken (1987) suggested that the fullest exploitation of vertical (height) as well as horizontal distribution of space was found in dooryard gardens, with their multiple levels of useful plants. He described such a garden in Mexico with a four-tiered vertical arrangement containing over 24 plant species. Similar home garden stratification is often found in the Latin American tropics (Denevan and Treacy 1987, Gliessman 1988).

There were 12 Nahuatl terms for different types of home gardens (Gonzalez 1985), indicating a comprehensive, broad knowledge and use of home gardens by the Nahuatl-speaking Aztec Indians of Mexico. The use of home gardens by present day Maya Indians has been reported in many parts of Central America and Mexico, and Wilken (1971) cited several references suggesting that home gardens were commonly planted by the ancient Maya. Ewell and Poleman (1980) described the agriculture of the Chinantec Indians near Tuxtapec, Mexico, where in one garden ninety edible plants were identified. In addition, 18 trees for wood, 6 for furniture and tools, and 55 plants for medicine were found.

Geertz (1963), in a much-cited paper, suggested that swidden garden plots mimicked the tropical forest, especially in their high degree of diversity. A number of authors (Beckerman 1983a, 1983b, Boster 1983, Hames 1983, Vickers 1983, Stocks 1983) have disagreed with some aspects of Geertz's model and reevaluated his concepts. They suggested that the degree of diversity suggested by Geertz is not found in most traditional systems studied. Weinstock (1984) noted that the studies of the above authors were primarily in South American forests where the primary crops were root and tuber crops such as cassava. In Asia, where Geertz did his studies, various grains such as rice were most important. Diversity would probably be much more significant in influencing disease development in grain crops, as compared to asexually propagated crops such as cassava in which diversity is primarily intraspecific rather then interspecific.

Beckerman (1983b) noted that the Bari or Motilones Bravos, Indians of Northern Colombia and Venezuela, often planted fields or household gardens near their houses. The fields nearest their longhouses received food scraps, garbage, ashes from fires, and human waste. Crops of high value were planted in these fertile gardens. Fields generally were planted in concentric rings around the central houses by the Bari. Gardens close to home sites in Africa also received household refuse and organic matter such as composts or animal manure (Miracle 1967).

Multistory Systems in Africa

The Chagga, a Bantu group living on Mt. Kilimanjaro in Tanzania are skilled traditional farmers who make use of multi-story gardens to support their dense populations (Fernandes et al. 1984). Their gardens included food and cash crops plus animals. More than 100 different plant species were recorded in Chagga home gardens. This number included 15 different types of bananas grown for food, brewing beer, and fodder. Teri and Mohamed (1988) reported that 40 different types of bananas are grown by the Chagga. Vertically, Fernandes et al. (1984) distinguished five relatively distinct zones in the Chagga home gardens. The lowest (0-1 m) contained various food crops, herbs, and grasses. The second zone (1-2.5 m) was comprised of coffee and various small trees and shrubs. The third zone (2.5-5 m) Fernandes et al. called the banana zone. This canopy also included some fruit and fodder trees. Next came a 5-20 m zone or canopy that consisted of fuel and fodder trees. Finally, the fifth zone (15-30 m) consisted of a canopy of valuable timber, fuel, and fodder trees. There was considerable overlap among these different zones.

Other Multistory Systems

A multistory cropping system developed in Papua New Guinea started out as a mixed vegetable garden and then was slowly converted into a coffee/banana and finally a coffee/casuarina system (Bourke 1985). Crops used in the system were bananas, cassava, taro, tannier, sugar cane, maize, and other local food crops. Farmers took into account the shading effects of the different crops, and eventually casuarina and coffee became the final stage of the system. The tree *Casuarina oligodon* fixes nitrogen, but its precise contribution to the system is not known.

Lagemann and Heuveldop (1983) described the coffee and shade tree cropping system found in Costa Rica. It is similar in some respects to local tropical forest ecosystems. Coffee was grown in association with at least one shade tree on 95% of the farms they surveyed in their project area (Acosta-Puriscal). The typical structure of the systems consisted of three layers: l) the lowest layer was coffee; 2) the second layer consisted of trees like *Inga* spp., fruit trees, or the nitrogen fixing *Erythrina poeppigiana*; 3) the third layer consisted of tall timber, fruit, or palm species. Clement (1986) described the utilization of the peach or pejibaye palm as shade in coffee plantations in Costa Rica.

Effect of Shade and Density in Multistory Systems

The density of crop cover, especially that of tropical trees, has other important effects on disease incidence. Palti (1981) cited Waller's description of how the density of tree foliage can effect tropical plant diseases:

> In tropical plantation crops, density of plant cover may have a twofold effect. In the rainy season, when rain runs down limbs and trunks, wet soil and foliage will take longer to dry under dense cover, and prolonged periods will favour many diseases, such as the coffee berry disease (*Colletotrichum coffeanum*). Conversely, in seasons poor in rain but rich in dew, dense plant cover will shield lower organs from dew formation, and will thus reduce the proportion of shoot growth in danger of attack by pathogens requiring films of water for their development.

The above effects, described for tropical plantation tree crops, would be similar for diseases in home gardens or tropical tree or shrub plantations.

Shading reduces the severity of several coffee diseases. Sridhar and Subramanian (1969), Nataraj and Subramanian (1975), and Wellman

(1972) reported fewer lesions due to brown leaf spot of coffee (*Cercospora coffeicola)* on shaded coffee. I have noticed fewer brown leaf spot lesions on coffee in Costa Rica under shade, as compared to coffee in full sunlight. Wellman (1972) also found fewer lesions due to American leaf spot of coffee (*Mycena citricolor*), less algal attack (*Cephaleuros virescens*), and less pink disease (*Corticium salmonicolor*) on shaded coffee. However, shade can also increase the severity of some diseases. Shade effects on disease are described in greater detail in Chapter 22.

Summary

There is little information in the literature on disease occurrence in the multistory system, but some reflection helps to explain why these systems often existed for centuries without major disease problems. First, traditional farmers for centuries selected landraces that could thrive in the conditions under which they were grown. A great diversity of crops was grown, and this provided a degree of protection because pests and pathogens are less able to build up to destructive proportions on the few isolated plants of each species. Various forms of polycropping also tended to reduce the seriousness of plant pathogens. The architecture of the individual plant was also taken into consideration by traditional farmers, especially in intercropping situations. For example, a cassava plant with a spreading habit would be unacceptable in an intercrop with maize as it would shade out the maize plants, whereas a tall, erect plant architecture would be desirable. Shade can have important effects on humidity, dew deposition, and temperatures, and these factors may reduce the severity of some pathogens. Plants under shade are also under less water stress than those in full sunlight (Allen et al. 1976).

Writing on the design of successful cropping systems in the tropics, Hart (1980) advised that the structure and function of natural systems in the same environment be studied and compared with cropping systems. A study of existing traditional cropping systems and their architecture also was strongly urged in the design or improvement of tropical cropping strategies. These multistory systems may provide an ideal model for other areas in the tropics.

21

Sanitation

According to dictionaries, sanitation means bringing about healthful conditions. As used in plant pathology, sanitation is a very broad term, and is used to describe a wide variety of practices. Generally, sanitation practices for the management of plant diseases include the removal or destruction of plants or plant parts that may be sources of inoculum for further disease spread. They may also include measures to prevent infection, such as the pruning or removal of healthy plant parts. There are overlapping boundaries among the practices described in this chapter and many others in this book.

Sanitation Practices

Palti (1981) wrote: "The two aims of sanitation are to prevent the introduction of inoculum into field, farm, or community, and to reduce or eliminate inoculum from diseased fields." This broad concept of sanitation would include countermeasures for all the means by which inoculum might enter a field or region. He listed seeds, propagating material, water, animals, plant debris, soil, compost, manure, farm implements and equipment, boxes and packing material, and humans as possible agents by which pathogens can be carried from place to place.

Sanitation practices include the destruction of plant residues, that carry pathogens, by burning, plowing under, or burying. Such field sanitation measures are recommended as management measures for many diseases of importance in the tropics (Kranz et al. 1977). An example is the elimination of potato dump piles and deep plowing of fields with tubers infected by *Phytophthora infestans* (late blight of potatoes). These practices reduce or eliminate the inoculum of *P. infestans* available the following year. Sanitation also includes preventing pathogen transmission by sanitary handling practices and disinfesting agricultural implements that might harbor pathogens.

Sanitation may be quite effective when a pathogen is localized on a plant part such as a leaf, twig, mummified fruit, or stem. Pruning of diseased parts has been recommended for the management of many diseases, such as crown gall of small fruits (*Agrobacterium* spp.), fire blight of apples and pears (*Erwinia amylovora*), black knot of *Prunus* spp. (*Dibotryon morbosum*), cedar apple rust (*Gymnosporangium juniperi-virginianae*), and witches' broom of cacao (*Crinipellis perniciosa*) (Stakman and Harrar 1957). Surgical practices such as removal of tree cankers often have been recommended. Detailed knowledge of the etiology and epidemiology of pathogens is essential for effective sanitation, as practices effective for one pathogen may be useless for another.

Almost all traditional potato growers in the Andes of South America plant whole seed (tubers) rather than cut seed (Figure 8.2), which is commonly used in the United States. It is well known that cutting potato seed is an excellent way to spread pathogens (especially bacteria and viruses), but it is possible to use cut seed in the US because of rigorous seed certification programs and strict sanitation practices. Nevertheless, serious problems due to the use of cut seed still cause serious losses in the US. As described in Chapter 1, when attempts were made by researchers to cut potato seed in Colombia, losses from soilborne pathogens and from bacterial blight (*Pseudomonas solanacearum*) were severe (Thurston 1963). The Colombian research program finally returned to using a practice traditional farmers knew was practical for their conditions. Colombian farmers probably had discovered over the centuries that cut seed would not produce a crop. Researchers had to rediscover a sanitation practice, that the traditional peasant farmers of Colombia already knew.

Yen (1974a) described the sanitation practice farmers in Rarotonga, Cook Islands use to avoid losses to sweet potato from *Cercospora* spp., which is prevalent and destructive on the island. The fungus attacks leaves, but by planting roots, rather than stem cuttings, which are most commonly used for propagation of sweet potatoes in the tropics, serious disease losses were prevented.

Roguing

Roguing is a sanitation practice that consists of the removal, from a crop, of individual plants (rogues) that are inferior or diseased. In addition to diseased plants, volunteers, mixtures, and off-type plants are often rogued from valuable crops or crops destined for seed. The practice has been widely used in plant disease management (De Bokx and van der Want 1987, Kranz et al. 1977, Palti 1981, Shepard and Claflin 1975, Stevens 1960, Stakman and Harrar 1957). Roguing is most often used when a crop is very valuable and the amount of infection is small. Most roguing is done by hand. Historically, roguing has been used for diseases of many crops, such as potatoes, tomatoes, cassava, sugar cane, maize, cacao, strawberries, and bananas. Viral, fungal, and bacterial diseases often are managed by roguing. According to Stevens (1960) and Stakman and Harrar (1957), roguing has been widely recommended but has rarely achieved a satisfactory level of control.

Roguing of virus-infected plants is commonly used in the production of foundation or certified seed potatoes (De Bokx and van der Want 1987). Fields many be rogued three or more times in a season in the production of foundation potato seed (Shepard and Claflin 1975). Most foundation and certified seed production regulations place a limit on the number or percentage of diseased plants allowed in a crop. Where viruses are seedborne, roguing is especially important in reducing sources of potential inoculum in the field. Traditional Aymara Indians of Peru were reported to rogue diseased plants from their potato fields (Huapaya 1982).

Williams (1984) described roguing as a useful and common practice for the management of most cereal downy mildews in Asia, and suggested that the practice reduced spread of asexual spores and production of oospores, which initiate infection in subsequent crops. Kenneth (1977) recommended roguing of infected pearl millet plants before oospores of *Sclerospora graminicola* (downy mildew) could form, as oospores are the primary source of inoculum for overseasoning.

Bock and Guthrie (1977) recommended the propagation of healthy cassava cuttings from plots that were strictly rogued as a management measure for African cassava mosaic.

Removing Infected Plant Parts

Williams (1981) noted that in China the practice of removing all leafy plant parts from the field immediately after harvest greatly reduces the potential for infected leaves to serve as sources of inoculum. He noted that the leaves removed are either fed to animals or composted.

Putter (1980) wrote that under experimental conditions taro blight (*Phytophthora colocasiae*) could be "turned on and off" simply by removing infected leaves daily for two days after the initial establishment of the disease, or by not removing them. Rosado May and

Garcia-Espinosa (1986) reported that in combatting web blight of beans Mexican farmers removed bean leaves infected with *Thanatephorus cucumeris* by hand in order to reduce fungal inoculum. Hand picking infected leaves for disease management is seldom practiced except in very small, isolated fields where plentiful labor is available.

In the 1960s, in Colombia, we received a report that tomatoes in the Cauca Valley were badly infected with viruses. After investigating, we found a high percentage of the tomatoes to be infected with tobacco mosaic. We also found that the male workers transplanting tomatoes smoked cigarettes infected with tobacco mosaic virus that is easily transmitted mechanically. Thus, smokers touching tomatoes for transplant were spreading the virus. Subsequently, when only non-smokers (e.g. women and children) were allowed to transplant tomatoes, infection by the virus was immediately reduced. This simple sanitation practice was able to bring about effective management of a serious disease.

Wellman (1938) suggested that the extreme poverty of some traditional farmers may have led to such a high level of sanitation that plant disease was effectively controlled. He visited extremely poor areas in Turkey in 1936 and found vegetables to be almost free of diseases, although wild relatives and escaped plants in the same areas were diseased. He attributed the disease control to the people's need to utilize every leaf and scrap of plant material (infected and non-infected) for human and animal food. Wellman wrote that in more prosperous areas, where it was not necessary to utilize every scrap of food, vegetables had the expected amount of disease.

Moniliophthora pod rot of cacao (*Moniliophthora roreri*) is a serious disease occurring in several tropical regions of Latin America (Thurston 1984). Sanitation practices can bring cacao plantings that have been abandoned due to *M. roreri* back into production. Barros (1966) recommended the following practices for the management of the disease: 1) elimination of suckers, 2) two prunings a year, 3) wound treatment with chemicals, 4) monthly harvests, 5) removal of diseased pods at each harvest, 6) improved drainage canals, and 7) three weedings per year. Similar cultural practices also were recommended for management of *Phytophthora palmivora* (black pod rot of cacao) by Asare-Nyako (1977). Although these measures were effective in reducing the losses due to the disease, many growers did not properly dispose of diseased pods and continued to suffer losses. Galindo (1987) and Porras et al. (1990) reported that cutting off infected pods before they sporulated and allowing them to fall to the ground where they quickly rot, was an effective disease management measure for *M. roreri* in Costa Rica.

Esca of Grapes

A disease of grapes known as esca (yesca in Spanish), caused by the fungi *Stereum hirsutum* and *Phellinus igniarius,* has been a serious problem of grapes in the Mediterranean area, apparently for millennia. The fungi cause a rot of the trunk and branches. Aristotle (384-323 B.C.) wrote: "A tree that has a hard bark and has become barren, if its roots be split and a stone inserted in the cleft, it will become fruitful again" (Ross 1913 as cited by Orlob 1973). In 60 A.D. Columela (1988), a Roman who lived in Southern Spain, described a surgical method of management for affected grape vines, namely splitting the trunk of the vine down the middle, scraping out infected tissue, and inserting soil dampened with amurca. Although no references were cited, Ruiz Castro (1944) wrote that a similar method, still used today in Spain, has been used since "time immemorial" in Greece and Asia Minor. Palladius, a Roman who lived in the fourth or fifth century, suggested driving a wedge into the roots to cure unhealthy grapes, olives, figs, palms, and peaches (Orlob 1973). A similar surgical practice for chestnuts and other woody plants was described by Ibn al-Awam (1988), a Spanish Moor writing in the 1200s. About 1350 the German Gottfried van Franken wrote that in Greece when grape vines began to deteriorate the roots were split, a rock inserted, and they were then covered with soil and manure (Orlob 1973).

It is highly probable that these management practices for the esca disease of grapes were adaptations of Aristotle's recommendation. Today in Spain the practice (Dominguez Garcia-Tejero 1989) is described as splitting the trunk down the middle with a hatchet, cleaning out (removing) infected tissue, and inserting a stone to keep the trunk separated. The resulting aeration appears to be effective in preventing further spread of the fungi. Bellod (1947) describes a similar practice, but doubts the value of using rocks. The lack of fallow and the planting of grapes in areas previously used for grapes were considered to be major reasons for the prevalence and severity of the esca disease, according to Boutelou (1949), in a book published in 1807 in Spain. Perhaps the insertion of soil into the cleft encouraged biological control of the esca pathogen by suppressive organisms. The practice is still used in Andalusia, Spain and is considered by some growers there to be an effective treatment for the esca disease of grapes (personal communication -- Antonio Trapero Casas). The alternative to the practice is the use of an arsenical fungicide. In 1990, I saw the practice being used near Montilla, Spain. Without understanding the many centuries of experience behind the above practice, one might ridicule it and consider it useless, whereas in fact the practice appears to be effective and useful.

Removing Healthy Plant Parts

Pruning removes plant parts, but does not necessarily remove infected plant parts. Although used primarily to increase fruit quality, pruning vineyards and orchards is common worldwide. It has been practiced since Biblical times (Rotem 1982). Thinned foliage allows better aeration and results in reduced humidity, quicker drying, better ventilation, and light penetration. Diseases generally are reduced in such thinned plantings. In 1348 the Spanish Muslim writer Ibn Lujun wrote that pruning benefitted small fruit trees, but added that pruning large trees could introduce diseases into them (Equaras Ibañez 1988).

Trutmann et al. (in press) describe the management practices of traditional farmers in the highlands of Zaire near Rwanda. Farmers remove leaves to alter the bean microenvironment; they even have a word (*gusoroma*) for the practice. Several reasons were given by farmers for leaf defoliation, including shade reduction and preventing neighboring plants from touching. One farmer stated: "If there is too much rain, we pick off the leaves of the beans to allow light in between the plants." These practices may affect disease incidence.

Varro (1934), a Roman who lived from 116-27 B.C., wrote the following regarding the harvesting of olives: "Those which cannot be reached with the hand should be beaten down; but a reed should be used rather than a pole, as the heavier blow renders necessary the work of the tree doctor." In Latin the last phrase reads: "*gravior enim plaga medicum quaerit*." It is tempting to read more into the term "tree doctor" than it probably deserves -- to interpret it to mean that injuries caused by poor harvesting practices were taken care of by a specialized individual (tree surgeon) who pruned off the injured tissues.

Removing plant parts does not always contribute to reduced disease. The young leaves and shoots of cassava are consumed as a leafy vegetable in several countries of tropical Africa and in a few Asian countries such as the Philippines. In Zaire people prefer leaves infected with African cassava mosaic and consider them to be sweeter and to have a softer texture (Almazan and Theberge 1989). Although the geminivirus is vectored by a whitefly, the frequent harvesting of leaves contributed to a high incidence of African cassava mosaic.

Management of Moko Disease by Sanitation

The bacterium *Pseudomonas solanacearum* probably attacks over 300 plant species and causes serious diseases of many important tropical crops. Moko disease, caused by one particular race of the bacterium (race 2), has caused enormous losses to bananas and plantains around the world. An insect-transmitted strain of the bacterium has been especially devastating on plantains in Latin America (Buddenhagen and Elasser 1962). Management of *P. solanacearum* is obtained with a combination of sanitation and removal of the male inflorescence (Buddenhagen and Sequeira 1958, Lozano et al. 1969). Insects (bees, wasps, moths, etc.) are an important means of dissemination of the pathogen and commonly frequent the flower. The source of the bacterium for insect transmission comes primarily from the male part of the banana flower, which may ooze bacteria for three months. When infected bracts of the male flower abscise, bacteria ooze from the bases of bracts and insects feed on them. If the male inflorescence is broken off by hand before the male flowers and bracts abscise, or is cut off by a properly disinfested machete, excellent control is obtained. The male flower plays no useful function. In an experiment in the Tolima Valley of Colombia, 328 racemes were harvested from 300 plants in plots where sanitation was practiced (10 months after the inoculation of plants in the center of the plots), but only 46 racemes were collected from 300 plants in plots where no control was practiced. The treatment was inexpensive, as in 1967 the cost was calculated to be only about $ 1.00 (US) per hectare (Lozano et al. 1969).

Although the above practice was not developed by traditional farmers, it illustrates a cheap, effective plant disease management practice that can spread rapidly among local farmers. Male flowers are used as food in some parts of Asia, and male buds are removed in some areas of the tropics because growers believe yields will be improved by their removal (Simmonds 1966). These traditional practices may facilitate the use of the above sanitation practice.

Composting

Composting can be considered a sanitation practice as it usually kills pathogens in the crop residues and other organic materials composted. In addition, when incorporated into the soil, composts often aid in disease management. Hoitink and Fahy (1986) reviewed the literature on the use of composts for disease management.

Traditional farmers used composts extensively in the Americas (Wilken 1987) and Asia (Cook and Baker 1983, McCalla and Plucknett 1981). These composting activities undoubtedly reduced the inoculum of soilborne pathogens. About 60 A.D. the Roman Columela (1988) described composting manure in trenches and sunken areas. The Arab

Ibn Luyun (Equaras Ibañez 1988) described composting of weeds and straw in 1348 and added that all manure was improved by storing it for one year, or better for two. Composting of manure and construction of manure pits has been recommended by several ancient writers (Cato 1934, Columella 1988, Ibn Luyun 1988, Varro 1934). The Roman Cato (1934), who lived 234-149 B.C., wrote the following: "You may make a compost of straw, lupines, chaff, bean stalks, husks, and ilex and oak leaves." Composts and other organic soil amendments contribute to better soil structure and to soil-borne pathogen management, and provide other agronomic benefits.

Burning

Burning is one of the most common methods used by traditional farmers for destruction of pathogen inoculum. The reasons traditional farmers use fire are several, but destruction of pathogen inoculum is probably not consciously one of them. Caveness (1971, IITA 1976) is one of the few researchers who has done research in this area, and his work was concentrated on nematodes. He found that burning a 10 cm litter layer in a slash and burn situation destroyed nematodes to a depth of 9 cm. The amount, moisture content, and quality of litter are important in burning. In addition, the wind velocity, duration and intensity of the burn, and soil type also affect the temperatures achieved. The use of fire to manage plant pathogens in modern agriculture was reviewed by Hardison (1976), but little information on the conscious use of fire to manage plant pathogens by traditional farmers was found in the literature. The use of fire relative to plant disease management is discussed in detail in Chapter 10.

Soil Solarization

Soil solarization is a soil disinfestation practice using solar heating of soil. In Israel (Katan 1981, 1987) and California (Pullman et at. 1981), thin, transparent, polyethylene tarps are used to cover the soil and trap solar energy. The system bears some resemblance to other techniques for heating the soil by solar energy during summer seasons that have been practiced by traditional farmers for centuries (Katan 1981). Soil solarization is a technique that might be appropriate for the farmers of some developing countries. The practice is more fully discussed in Chapter 17.

Flooding

Another major means of reducing or eliminating inoculum in fields is by flooding. The paddy rice system is important in the management of soilborne diseases in rice-growing areas of Asia and Africa. Traditional farmers have used flooding for plant disease management for millennia. Over 220 million ha of cropland are irrigated worldwide (FAO 1986), and much of this is periodically flooded. The anaerobic or near anaerobic conditions produced by flooding are known to reduce populations of many soil pathogens (Baker and Cook 1974, Palti 1981, Stolzy and Sojka 1984). Flooding is further discussed in Chapter 11.

Summary

The term sanitation is used by plant pathologists to describe a variety of practices that result in a reduction of pathogen inoculum. Sanitation practices include the removal, roguing or destruction of plants or plant parts that may be sources of inoculum for further disease spread, or measures to prevent their introduction. Various sanitation practices used for disease management also are described in more detail in the other chapters. For example, sanitation practices are often an important component of biological control. Flooding, burning, composting, soil solarization, many tillage practices, and production of clean seed are examples of practices that result in sanitation and enhanced disease management.

Traditional farmers use a variety of practices that could be classified as sanitation, but usually with objectives in mind other than managing plant diseases. Some practices, such as flooding and burning, may significantly reduce pathogen inoculum, but traditional farmers use the practices primarily for other reasons. Some effective labor-intensive sanitation practices may be feasible for traditional farmers, whereas labor costs in modern agriculture might make such practices prohibitively expensive.

22

Shade

Manipulation of shade is seldom considered a plant disease management practice, but for some crops, especially tropical crops, management of shade or light is important in determining yields and disease severity. Shade was almost universally used for coffee until the latter half of this century, and some tea was also grown under shade. Fernandez de Oviedo (1986), writing in the 1600s, noted that Indians in Nicaragua used shade trees for cacao, and that they pruned the trees to give the proper shade. Careful management of the degree, quality, and source of shade is most important. Palti (1981) distinguishes three types of shade: 1) shade given by slopes, 2) shade given by vertical objects bordering the field or planting, and 3) shade given by horizontal objects. Shading by trees may produce either vertical or horizontal effects.

Effects of Shading

The effects of shading are highly complex (Willey 1975), and there is a need to better understand its action relative to plant disease. Shade trees affect not only light intensity and quality, but also air circulation, soil and air temperatures, soil moisture, and deposition of dew and rain. Shade trees also protect plants beneath the canopy from hail and strong winds. Wrigley (1988) states that in Indonesia temperatures in shaded coffee averaged 7°C lower than ambient temperatures during the day and 3°C higher at night. Organic and inorganic nutrients are added to the soil by processes of leaf deposition and decomposition. Competitive effects obviously occur also.

Poró (*Erythrina poeppigiana*) is a commonly used coffee shade tree in Costa Rica . Trees are pruned 1-3 times a year, and the pruned branches provide a mulch and return nutrients to the soil (Figure 22.1). It has been judged that this practice may return over 300 kg of N/hectare to the soil per year (Glover and Beer 1986, Russo and Budowski 1986). Bornemiza (1982) estimated that 100 kg N/ha/yr was sufficient for low to medium density plantings of coffee, but not for high density plantings (i.e. > 5000 plants/ha). More recently, Beer (1988) concluded that *E. poeppigiana*, when pruned 2-3 times per year, can return to the litter layer the same amount of nutrients applied to coffee plantations in Costa Rica via inorganic fertilizers, even at the highest recommended rates of 270 kg N, 60 kg P (phosphorus) and 150 kg K (potassium)/ha/yr. In addition, trees contribute 5000 to 6000 kg organic matter/ha/yr. Although the nutrient contribution is important, Beer concluded that in fertilized plantations of cacao and coffee, litter productivity is a more important shade-tree contribution than nitrogen fixation, due to the beneficial effects of all that organic matter on soil physical and chemical properties.

The term allelopathy, as defined by Holliday (1989), is used for both the harmful and beneficial effects of higher plants on the germination and growth of another plant. The effect "is exerted through the release of a chemical by the donor." In some cases the effects of allelopathy may have been confused by ancient writers with the effects of shade on crops. Varro (1934), a Roman who lived 116-27 B.C., wrote: "large numbers of large walnut trees (*Juglans* spp.) render the border of the farm sterile." Pliny (cited by Rice 1984) wrote in 1 A.D. that the walnut was injurious to anything planted in its vicinity. Ibn al-Awam (1988), a Spanish Arab, cited the Islamic author Háj in the 1200s as stating that the walnut has an aversion to most other trees ("Dice Háj, que el nogal aborrece la cercanía de la mayor parte de los arboles"). The Spanish writer Alonso de Herrera (1988) wrote in 1513 that the shade of walnut trees damaged other trees and plants and added that walnut leaves should be carefully disposed of; otherwise seed planted in the areas where leaves fell would be damaged. Alonso de Herrera's observations on damage were correct, but the problems walnuts caused were probably allelopathic effects and not shade effects.

Advantages and Disadvantages of Shading

The dense shade produced by the canopy of some plants is important in controlling weeds. For example, cassava produces a dense canopy with intense shade, and some traditional farmers take advantage of this by weeding only until the canopy discourages further weed growth. (Figure 22.2) This weed control is of considerable importance in hot, humid areas where weed growth is otherwise abundant and rapid. This practice was noted in the Caribbean islands by Fernandez de Oviedo

(1986) in 1526 who wrote: "And thus as the cassava continues to grow, and the weeds are controlled, until such time as the cassava dominates (shades) the weeds." When cassava plants are severely defoliated by disease, weeds may begin to grow again under the thinned cassava canopy, and yield losses due to weeds may be greater than the effects of the foliar disease.

Beer (1987) reviewed the advantages and disadvantages of shade for coffee and cacao, and Alvim (1977) and Smith (1985) did the same for cacao. Since approximately 1950, growing coffee and tea in full sunlight has become common in many tropical countries and is frequently recommended. Striking increases in yield have been obtained by many investigators when coffee and cacao were grown without shade (Alvim 1977, Ahenkorah et al. 1974, Purseglove 1968).

However, there may be long-term disadvantages to the practice of growing coffee and cacao in full sunlight. Ahenkorah et al. (1974) summarized the results of a 17 year shade and fertilizer trial in Ghana. Cacao fertilized and grown without shade yielded three times as much as shaded trees. However, cacao without shade was more susceptible to insects and, according to the authors, "probably" to diseases. Furthermore, after ten years the unshaded trees declined in vigor, due to a high incidence of pests, establishment of mistletoe (*Tapinathus bangwensis*) and mosses, loss of exchangeable bases from the soil , and depletion of reserves. Alvim (1977) suggested that lack of protection from the wind may have been an additional factor. Regarding Ahenkorah's results Alvim wrote:

> The above findings are in agreement with practical observations by cacao farmers in many countries showing that increase in yield following complete removal of shade usually does not last very long and is followed by rapid decline of plantations, with many plants showing severe defoliation and dieback after the third or fourth year.

Bigger (1981) studied insect populations of shaded and unshaded Amelonado cocoa. He found larger populations of mealy bugs (*Planococcoides njalensis*) in cacao grown in full sunlight. Since the mealy bug is the vector of cacao swollen shoot virus, which causes a serious disease in West Africa, perhaps growing cacao in full sunlight might also result in more disease.

In practice some degree of partial shade is probably needed for cacao plantings. The proper degree of shade may not always be easy to obtain, considering the multitude of conditions and objectives under which these multiple cropping systems are managed, varietal differences, soil differences, the large diversity of shade trees utilized, and the different climates where cacao is grown. Purseglove (1968) states: " In its natural habitat *Theobroma cacao* is a small tree in the lowest story of the evergreen tropical rain-forest of South America." Traditional farmers in tropical Latin America have cultivated cacao, probably almost always under shade, for over 2,000 years (Purseglove 1968). The long-term effects of growing cacao and other shade-tolerant crop species in full sunlight need to be carefully studied in long-term experiments such as those described by Ahenkorah et al. (1974). Beer (1987) suggested that in the absence of intensive management, the use of shade should be continued for cacao and coffee.

Increased Disease Under Shade

Some diseases cause greater losses under shade. Black pod rot of cacao, caused by *Phytophthora palmivora*, is reported to be more severe under shade (Besse 1972, Dakwa 1979). Smith (1985) noted that four of the shade trees recommended for cacao in Papua New Guinea were susceptible to two cacao pathogens (*P. palmivora* and *Corticium salmonicolor*), and thus may have served as sources of inoculum when used as shade. Asomaning and Lockard (1964) reported suppression of cacao swollen shoot symptoms by light. Blister blight of tea, caused by *Exobasidium vexans*, is a destructive disease in Asia and was more severe in shaded than in unshaded tea (Kerr and Rodrigo 1967, Visser et al. 1961). Schieber (1977) recommended reducing shade for the management of *Mycena citricolor* (American leaf spot of coffee) and *Ceratocystis fimbriata* (*mal de machete* of cacao). The high humidity, lack of air movement, and reduced light under shade increases the severity and incidence of some diseases.

Decreased Disease Under Shade

Some diseases are suppressed by shade. Thorold noted in 1940 that bananas grown under shade in cacao plantations had fewer lesions caused by *Mycosphaerella musicola* (Sigatoka disease) than those grown in full sunlight. He also noted that bananas on steep slopes had less damage due to Sigatoka than bananas grown in level areas, because they had fewer hours of direct sunlight. According to Thorold (1940), moderate shading of bananas was even recommended as a disease management measure. Carment (1922), Maas (1967), and Wellman (1972) reported fewer Cercospora lesions on shaded bananas than on bananas in full sunlight. Stover (1987) noted fewer Black Sigatoka (*Mycosphaerella fijiiensis*) lesions on plantains shaded under coconuts than on plantains in full sunlight. In traveling in the tropics, I too have frequently noted fewer lesions caused by *M. musae* on shaded bananas than on bananas in full sunlight.

According to Karani (1986), for centuries in Uganda bananas were grown in association with large trees (*Ficus* spp. and *Ficus natalensis*) that provided shade. The trees were pruned to give sunlight in the morning and afternoon, and bananas and other crops were grown beneath them. Bark cloth for clothing, which was later displaced by cotton cloth, was made from the trees. During the last 40 years, most of the trees have been eliminated, and the importance of *Mycosphaerella musicola* and other pests has increased, probably due to the elimination of shade from this sustainable agroforestry system.

A similar situation exists with brown leaf spot of coffee caused by *Cercospora coffeicola*, as the disease is less serious on shaded coffee than on coffee grown without shade. Several authors (Nataraj and Subramanian 1975, Sridhar and Subramanian 1969, Waller 1977, Wellman 1972) reported fewer brown leaf spot lesions on shaded coffee. On numerous occasions I have also noticed less brown leaf spot on coffee in Central America under shade as compared to coffee in full sunlight.

Wellman (1972) found less American leaf spot of coffee (*Mycena citricolor*), less algal attack (*Cephaleuros virescens*), and less pink disease (*Corticium salmonicolor*) on shaded coffee. Wrigley (1988) reported that Elgon die-back (*Pseudomonas syringae* pv. *garcae*), which occurs at high altitudes in Kenya and Uganda, was reduced by growing coffee under shade. In Tanzania, Tapley (1961) described a problem of unknown etiology called "crinkle leaf", the symptoms of which were mild in shaded coffee and severe in unshaded coffee.

Wellman (1972) found less angular leaf spot of bean (*Phaeoisariopsis griseola*) where beans were shaded by corn stalks in intercropping situations. He also found less citrus withertip (*Glomerella cingulata*) and citrus scab (*Elsinoe fawcettii*) in the shade. Adegbola (1977), working in Nigeria, found symptoms caused by cacao swollen shoot virus more severe in full sunlight than in shade. Fagan (1987) noted no infection of coconuts by *Drechslera incurvata* (Drechslera leaf spot of coconuts) under the dry conditions of a greenhouse. Experimental shading of coconut seedlings in the field (30% or 50% shade with shade cloth) depressed dew formation and reduced the number of infections and their subsequent severity. Hilton (1952) reported from Malaysia that shaded rubber plants were relatively free of rubber bird's eye spot (*Drechslera heveae*), and that lesions of the fungus sporulated 85% less under shade than in the open.

Summary

I have found scant information to explain why disease is more serious or less so under shade. Palti (1981) notes that effects on rain distribution are pronounced in tree crops with dense crowns as rain drains off unevenly. Dense crowns may exacerbate soil splashing and thus increase the damage by soilborne pathogens such as *Phytophthora citrophthora*, which causes brown rot and gummosis of citrus fruits and trees. Palti also suggests that, with tropical plantation crops, dense plant cover (shade) may cause soil and foliage to remain wet for long periods, thus favoring some diseases such as the coffee berry disease (*Colletotrichum coffeanum*). A different effect may occur in drier seasons, as dew forms only infrequently, if at all, under dense shade. Thus pathogens that require free water for infection may not have sufficient hours of dew for infection. Air circulation is less under shade, and this may have an effect on some pathogens. Referring to tree diseases, Yarwood (1951) observed that anthracnose diseases are favored by rain, rusts by dew, and powdery mildews by shade and dry weather. From Yarwood's observations, we can judge that rusts would be less serious under shade as wetting and dew deposition would be less, whereas anthracnose diseases would be enhanced because of prolonged wetting after rains.

Coffee without shade tends to overbear and, if fertilizers are not adequate, will exhaust itself. Coffee is then more susceptible to many diseases, as are most weakened plants. Shade not only controls weed growth, it also changes species occurrence; broadleaf weeds tend to appear under shade, grasses in the open. As grasses compete much more strongly than the broadleaf weeds (at least under the conditions found in Costa Rice coffee plantations), this is another advantage of shade. Greater competition could lead to weakened crops and more disease. Shade effects are extremely complex and the explanation of greater disease with or without shade will rarely be a direct interaction (personal communication -- J. Beer).

It is probable that traditional farmers probably know from long experience crops which can be grown successfully in full sunlight and which cannot. Farmers in Costa Rica suggest that the fungal disease "ojo de gallo" (*Mycena citicolor*) is more severe under tall shade trees due to drip damage (personal communication -- J. Beer). Productive household gardens in the tropics, which utilize multistory cropping, are examples of traditional shading manipulations that often result in effective disease management. The use of shade should be added to possible management practices to be evaluated more fully for their usefulness in the tropics.

23

Weeds and Wild Plants

Many traditional farmers spend an inordinate amount of time in weed control (Figure 23.1). Kasasian (1971) suggested that more human effort is devoted to weed control than any other single human activity. Farmers in Nigeria may spend up to half of their time in weed control (Moody 1975). Weeds are a major cause of the abandonment of slash and burn plots after a few years (Eden and Andrade 1987, Sanchez and Benites 1987). According to Hammerton (1985), East Caribbean farmers spend about 30% of their time on post-planting weed control, most of it by hand. De Datta and Lacsina (1972) found that if weeds in rice were uncontrolled, they could reduce production up to 80%. Weeds are especially important in the humid tropics where they are most difficult to manage (Akobundo 1980, Kasasian 1971).

Weeds can be beneficial. For example, the leaves of innumerable species of "weeds" are used as pot-herbs in Latin America, Asia, and Africa (Brokenshaw and Riley 1980, Brown and Marten 1986, Gade 1975, Martin and Ruberte 1975, Oomen and Grubben 1978, Rocheleau et al. 1990). Gade (1975) described the use of weeds in Peru as a source of food for humans and animals. Efraim Hernandez X. (personal communication) noted that about 40 of the weed species found in Mexico's corn fields were also eaten as pot herbs by farmers. Brokenshaw and Riley (1980) described the use of wild plants in Kenya by the Mbeere as follows: "Many wild plants are eaten in Mbeere in various forms; leaves are used as vegetables, to add to the main meal; wild fruits are eaten as snacks; and a great variety of edible plants are used in times of famine." A study of women's use of wild plants in Kenya found that 65 species were used for food and 99 for medicine (Rocheleau et al. 1990). Johns and Kokwaro (1991) describe the many wild plants used for food and medicine by the of the Luo in Kenya. Brown and Marten (1986) and Gade (1975) listed additional positive uses of weeds in traditional societies, including their value as mulches, medicine, firewood, thatch, and food for animals.

Weeds, Wild Plants, and Plant Diseases

The role of weeds relative to plant diseases has been reviewed by Bos (1981), Duffus (1971), Hooper and Stone (1981), Palti (1981), and Thresh (1981a). Weeds may be hosts to all classes of pathogens, and are one of the most important means whereby pathogens spread and survive adverse conditions. Palti (1981) suggested that weed management should be an integral part of disease management and gave examples of pathogens that attacked literally hundreds of hosts, including weeds. For example, cucumber mosaic virus, which infects many crops of importance to traditional farmers, such as bananas, various cucurbits, beans, tomatoes, peas, and crucifiers, has at least 775 hosts in 85 different plant families. The virus is also known to be seedborne in some important weeds and is readily spread to a susceptible crop. *Phymatotrichum omnivorum*, a fungus causing root rot of cotton, infects 1,300 species of plants. Root-knot nematodes (*Meloidogyne* spp.) infect over 700 host species, including many weeds (Palti 1981).

Many vector-borne viruses spread through weed hosts. Some, such as barley yellow dwarf virus, cowpea mosaic virus, various maize viruses, and rice tungro virus, may cause serious losses to crops of traditional farmers. Thresh (1981b) noted that the distribution of weeds relative to the host may be important. Weeds may be situated within the crop, alongside the crop, or a long distance from the crop. The distribution of weeds has an important effect on virus spread by vectors. Allen (1983) gave a number of examples of weeds and wild plants that may play an important role in the ecology of legume viruses by serving as reservoirs of vectors for viruses.

Palti (1981) lists various air-borne pathogens known to spread through weeds and wild hosts, such as the pathogens causing stem and stripe rust of wheat and the various powdery mildews of cereals and cucurbits. Wild grasses are especially important in the evolution and spread of the cereal rusts (Browning 1975, Wahl et al. 1984). The alternate hosts of several important cereal rusts are wild plants or weeds, and their eradication in the past has been an important disease management practice. For example, the alternate host of black stem rust of wheat (*Puccinia graminis* f. sp. *tritici*) is the barberry. According to Stakman and Harrar (1957), farmers in the area of Rouen, France, instigated in 1660 the passage of a law requiring the destruction of barberries Obviously, by connecting the barberry with black stem rust, the farmers knew something that the scientists of the time did not. Campaigns beginning in the US in 1918 eradicated about one-half billion barberry bushes, primarily in the midwest (Stakman and Harrar 1957).

Some soilborne pathogens survive in the soil because of their ability to infect weeds. Root-knot nematodes (*Meloidogyne* spp.), *Verticillium* spp., and *Rhizoctonia solani* are examples listed by Palti (1981). Hooper and Stone (1981) have reviewed the literature on the role of weeds relative to nematodes.

Jones (1981) wrote that an enormous number of wild plants, both solanaceous and others not related to the potato, act as reservoirs of viruses and vectors for the many potato viruses found in the Andes of South America. Crops such as oca, olluco, and mashua are commonly interplanted with potatoes in the Andes. Oca has been shown to be a host of at least two potato viruses.

Mizukami and Wakimoto (1969) reported that weeds (*Leersia* spp.) growing near paddy fields were an important source of bacterial leaf blight of rice (*Xanthomonas campestris* pv. *oryzae*) inoculum. Padmanabhan (1977) recommended destruction of the weeds *Leersia hexandra* and *Echinochloa colona* near rice fields, as they are sources of inoculum for *Cochliobolus miyabeanus* (causal agent of brown spot of rice).

Profuse weed growth may increase the humidity within a crop canopy by reducing air circulation, and this increased humidity may favor some pathogens. The shade produced by weeds may have either a negative or a positive effect on the spread and severity of pathogens (discussed in Chapter 22). Eradication of weeds may be an important component of disease management, although in many cases eradication is impractical or non-economic.

Weeds and Traditional Farming in Mexico

The appearance of the maize fields of traditional farmers in Mexico might surprise most temperate zone farmers, as near harvest time their fields are usually choked with weeds. Mexican scientists have observed that farmers weed their fields for about ninety days, and then often let the weeds grow. Farmers have found that under their conditions, little additional yield increase results from additional weedings. Furthermore, weeds are used as fodder for animals in the dry season, and farmers have noted that there is far less wind and water erosion when weeds are present in a field. Efraim Hernandez X. (personal communication) noted that about 40 of the weed species found in Mexico's corn fields were also eaten as pot herbs by farmers. Some weeds were allowed to go to seed in order to encourage their growth. Thus, the weeds in Mexican farmer's fields are not necessarily there because of poor farming practices.

Chacon and Gliessman (1982) wrote that traditional farmers in Tabasco, Mexico, did not have a word for "weed" in their vocabulary, but instead used a concept of "good" and "bad" plants (*mal y buen monte*). Furthermore, the same plant could be either good or bad depending on where and when it was found. Thus, to Mexican traditional farmers, weeds are an integral part of their agriculture. Studies have shown that farmers had a better understanding of the agroecosystem than they were given credit for and have suggested that traditional farmers tried to manage weeds as a component of the agroecosystem rather than trying to eliminate them completely. Rosado-May and Gliessman (1988) and Rosado-May et al. (1986), working in the same area of Mexico, reported that beggar's tick (*Bidens pilosa*) was considered to be a "bad" weed by farmers, because it caused slow growth and chlorosis of maize. They suggested that its negative effects on maize were not due to pathogens (fungi or nematodes) but rather to allelopathy. They also found that *B. pilosa* suppressed nematodes.

Summary

Although traditional farmers may expend more labor in controlling weeds than any other single activity, weeds may also be perceived by traditional farmers to have value for food, mulches, medicine, firewood, thatch, and forage for animals. Weeds may serve as a reservoir of inoculum for various diseases and a means for pathogen inoculum to survive adverse conditions and overseason. Weeds often play an important role in the epidemiology of diseases. Weed growth may increase humidity and shade within a crop and thereby affect disease incidence. The role of weeds should always be considered in plant disease management

HOST PLANT RESISTANCE

24

Selection, Diversity, and Resistance

Scientific breeding of plants for disease resistance probably did not begin until after the disastrous potato late blight epidemic in Ireland during 1845 and 1846. An estimated million Irish died in the resulting famine. However, traditional farmers have been selecting for disease resistance for thousands of years. Peter Jennings (1976), who developed the rice variety IR8 , which began the green revolution in rice, stated: "The breeding methods devised by Neolithic Man remained standard until the 20th century, although in recent decades they were applied more systematically and with more sophistication. The technique is called pure line selection."

The ancient Romans carefully selected grain for future plantings. Columela (1988), writing in 60 A.D., suggested that the largest spikes of grain be selected each year for the next years' planting. He also cited Virgil (70-19 B.C.) as suggesting that cereals would degenerate if the largest seed was not selected for planting, one by one, each year. According to Stakman and Harrar (1957), the Romans accumulated and recorded information on the relative susceptibility of various crop species to rust , and such information certainly affected variety selection.

Vavilov (1951) was among the first to recognize that the centers of crop origin were also centers of genetic diversity. His "micro" concept of centers has been expanded to a "macro" concept including large regions, sometimes spreading in a band across huge areas of a continent (Harlan 1971). In these centers, the host, and usually a complex of various pests and pathogens, coevolved while these crops were being cultivated by traditional societies over centuries or millennia (Browning 1975). Landraces (traditional cultivars) with local adaptation were selected in these centers, under the prevailing peasant systems of agriculture. Buddenhagen (1981) wrote:

> With all the emphasis on breeding new varieties with superior quality, adaptability, disease and insect resistance, it is a shock to realize that millions of acres of many crops are varieties, or landraces, or clones selected by ancient men and women in prehistory, or at least before agricultural science was developed. This is so for the Dioscorea yams, for most of the rice and cowpeas in West Africa, and for most of the maize and beans in Latin America. It is still true for several million acres of rice in Asia, and much of the potato crop of the Andes in South America...... Sorghum and millet in tropical Africa are largely old landraces, as is much of the forage grass acreage of the world.

Few scientists would disagree with the premise that diversity is a useful characteristic for crop species and for natural and agricultural ecosystems. Natural plant communities rarely have the serious plant disease epidemics that occur frequently in the monocultures of our modern agriculture. These concepts have been extensively discussed by Brown (1983), Browning and Frey (1981), Brush (1986), Clawson (1985), Hawkes (1983), Murdoch (1975), Suneson (1960), and Vavilov (1951). Leppik (1970) wrote that the maximum number of genes for disease resistance are found in landraces, where host and pathogen have coexisted for long periods of time. Browning (1975) analyzed not only diversity in crops, but also diversity in their pathogens in their centers of origin, and suggested that there is more diversity, not only in crops, but also in their pathogens in these centers. Browning recommended the establishment of "living-gene parks" to maintain diversity not only in crop progenitors, but also in their pathogens. Altieri and Merrick (1987) suggested that crop genetic resources might be saved through maintenance of traditional farming systems. Genetic diversity has been defined by Brown (1983) as "the amount of genetic variability among individuals of a variety, population, or a species." The diversity found within the landraces selected and maintained by traditional farmers during millennia constitutes a tremendous contribution to the human race.

Nature's Diversity in the Tropics

Traditional agriculture, especially in the hot, humid tropics, is often highly diverse with regard to the number of species grown and the planting patterns and architecture of the agroecosystems used. Agricultural systems in the tropics are especially complex and often resemble the highly complex and diverse tropical forests of the region where they are found. The diversity and complexity of tropical ecosystems and agroecosystems are well documented by numerous investigators.

An illustration of such diversity is found in the Philippines. Mount Makiling is a rather small mountain in Luzon, Republic of the Philippines, some 1,125 m high and 8 km in basal diameter (Stevens 1932). Myers (1988a) stated: "The Mount Makiling area alone contains more woody species than the whole of the United States." Longman and Jenik (1987) wrote that in Southeast Asia it is common to find 100 different tree species per hectare of rain forest, and that occasionally as many as 400 woody species may be found per hectare. An additional illustration of the diversity of tropical forests was given by Myers (1988b): "In the central portion of the La Selva Forest Reserve in Costa Rica, totalling only 13.7 km^2, there are almost 1,500 plant species, more than in the whole of Great Britain with its 243,500 km^2." Myers also related: "Ecuador contains an estimated 20,000 plant species, the bulk of them in its forests and at least 4,000 of them endemic. By contrast, the temperate zone state of Minnesota, twice the size of Ecuador, has 1,700 plant species." Figure 24.1 illustrates a rain forest in South America, much of which is being lost to pastures.

Denslow (1988) related that Costa Rica, with an area about the size of West Virginia, supports more than 12,000 species of vascular plants and 850 species of birds. That is more bird species than the US and Canada combined. In 1964 there were 1556 species of birds recorded in Colombia, which is twice as many as the US can claim (Meyer de Schauensee 1964). Raven (1981) wrote that there are about as many species of plants in Panama as in all of Europe. Wilson (1988a, 1988b) stated that although tropical rain forests cover only 7% of the world's land surface, more than half of the worlds species are found in them. The rich diversity of the tropical forests gives them protection against diseases and other pests in a moist, warm environment ideal for the propagation of many plant pathogens. Relative to diseases in tropical forests Harlan (1976) has written:

> As a rule, they are immensely rich in species deployed, it seems for maximum protection against disease epidemics. Individual specimens of a given species are likely to be widely separated, rarely occurring in groves or colonies. The whole flora is such a heterogeneous mixture that it would appear difficult for any pathogen to build up sufficient inoculum to increase the population of lesions very much. Endemic diseases are numerous, but natural epidemics unlikely.

Examples abound of the diversity found in many tropical agroecosystems (Barker 1884, Boster 1983, Conklin 1954, Denevan and

Padoch 1987, Miracle 1967, Salik 1989). It should be noted that species diversity declines in the tropics with increasing altitude and increasing latitude.

Landraces and Centers of Genetic Diversity

Buddenhagen (1983) noted that landraces: " are adapted to the diverse physical stresses and biological problems of local areas where they are grown under low inputs and low densities, sometimes in mixed cropping systems." Harlan (1975) wrote:

> Landraces have a certain genetic integrity. They are recognizable morphologically; farmers have names for them and different landraces are understood to differ in adaptation to soil type, time of seeding, date of maturity, height, nutritive value, use and other properties. Most important, they are genetically diverse. Such balanced populations -- variable, in equilibrium with both environment and pathogens, and genetically dynamic, are our heritage from past generations and cultivators.

Harlan (1976) also described land race selection by traditional farmers:

> The techniques of subsistence agriculture select continuously for a quality the French call "rusticité" and for which there is no appropriate English word. Cultivars or landraces with rusticité yield something despite flood, drought, disease, insects, nematodes, birds, monkeys, or witchweed. In basic subsistence agriculture there is little pressure for high yields, but a crop failure means famine and starvation. Dependability is essential, a matter of life or death to the cultivator. Crops developed under these conditions are hardly ever high yielding, but they have excellent general fitness for local conditions and good wide-spectrum resistance to the diseases of the region. Specialized, derived cultivars developed by modern breeding methods may be high-yielding, but this is usually achieved at the cost of restricted and specialized fitness.

One center of diversity for potatoes is located around Lake Titicaca in Peru and Bolivia in the South American Andes, where today the greatest diversity of potato germplasm in the world is found. Resistance to almost every pathogen for which pathologists and breeders have searched has been identified in South American potato germplasm (Plaisted et al. 1987). Today, however, the potato landraces selected during thousands of years of cultivation by traditional farmers are being replaced by a few improved varieties developed by agricultural scientists. In Colombia, for example, traditional cultivars are rapidly being replaced by new varieties. Gomez and van de Zaag (1986) stated: "Currently, less than 15% of the potato area in Colombia is cropped with old cultivars and these persist mainly because of their outstanding cooking quality and taste." Brush (1986) related that 60% of the potato varieties in commercially developed valleys, such as the Mantaro Valley of Peru, are improved varieties, while away from these areas only about 25% are new varieties.

In the Andes traditional potato growers often grow a large number of potato cultivars in the same field (Rabinowitz et al. 1990). In addition, farmers may have a number of separate plots in different locations and at different altitudes, which have different environmental conditions. Mayer (1979) described an altitudinal distribution of potato cultivars in Peru. A farming community in Bolivia was described by Carter and Mamani (1983) in which the average traditional family owned 21 different plots in the potato-growing areas. Over 2,000 named potato cultivars were grown in Peru. In one isolated Peruvian valley with about 1,000 habitants, over 50 potato cultivars were found (Brush 1977). Brush et al. (1981) described the selection and maintainence of potato varieties in Peru by traditional farmers as follows:

> Botanists have suggested that native Andean potato fields are dynamic evolutionary systems. Inter-- and intra--ploidy hybridization is made possible by the planting practices of Andean farmers reported here and elsewhere. Our research indicates, however, that selection and distribution of native potato varieties is far more complex than the random planting of numerous varieties. The maintainence of these numerous varieties is neither casual nor random. A regular system of nomenclature, organized in a taxonomic manner, accompanies this cultivation. Specific cultivars are identified according to an implicit set of criteria involving tuber and sometimes other characteristics. Selection of varieties relates primarily to culinary characteristics, but specific biological fitness is recognized in certain clones. The adaptation of native varieties to local conditions is revealed in the difference between native and hybrid or improved varieties. The latter depend on frequent replenishment of "seed" to remain viable. The maintainence of clonal heterogeneity is a regular part of native agricultural practices, as is the distribution of a few cosmopolitan clones through "seed" networks. The cultural and biotic features of Andean potato agriculture are thus interwoven and express a complex symbiotic relationship between man and plant.

De Murúa (1987) wrote that when Inca farmers found a new type of potato, other root crop, or maize they would have a celebration that included singing, dancing, drinking, and a procession. Such activities would certainly indicate an extremely strong interest in plant selection and improvement.

Maize culture in Mexico provides another illustration of the selection practices of traditional farmers. Mexican farmers near Puebla, Mexico, have been growing maize for perhaps 7,000 years, and thus they have centuries of accumulated experience with the crop. Most maize varieties grown near Puebla are native landraces that have been selected and maintained over millennia by Mexican Indians, and they are better adapted to the area and perform better than those available from the government or the International Center for Improvement of Maize and Wheat (CIMMYT 1974). Also, maize is not grown as a monoculture, but rather is grown together with squash and climbing beans. Studies have shown that not only economic yields, but also nutritional yields are often superior with this cropping system. Criteria for selection of maize for this multiple cropping system would be far different from those used for selection of maize for a monoculture, such as is commonly found in the US cornbelt.

In a wonderful book entitled "*Indian Corn in Old America*," Weatherwax (1954) described the selection and maintainence of maize in Mexico and Central America by Indian farmers:

> As the ears are husked, they were often sorted into groups according to color and size; but, to retain the variation necessary for the game, the shelled grain of the different kinds was often mixed together again for planting. Unwittingly they were thus retaining the hybrid condition necessary for the maximum vigor of the plant.

Weatherwax noted that the Indians of the Americas had separated the major variety complexes of maize before the arrival of the Europeans. He related that one of the best taxonomic statements he ever heard on maize came from an illiterate Indian woman in a market in Mexico when asked how many kinds of maize her people had. She answered as follows:

> We have four months maize and six months maize; then there is white maize and black maize and yellow maize; and there is large maize and small maize. But they are all the same thing. There is just one name for all of them. They are all maize.

Interspecific Diversity

There are countless descriptions, especially in the anthropological literature, of the large numbers of species cultivated by slash and burn agriculturalists. Conklin's (1961) review article on slash and burn agriculture includes 1,200 references and discusses the complexity and diversity of slash and burn cropping systems. The Hanunóo people of the Philippines cultivated 413 different kinds of plants , 87 of which were basic plants grown in slash and burn fields (Conklin 1957). Generally 40-50 plant types were grown in the swiddens in the first year. This diversity of species provided a degree of protection because pests and pathogens are seldom able to build up to destructive proportions on the few isolated plants of each species. Individual species within a tropical forest are generally rather uniformly dispersed, and this strategy is also useful in preventing disease epidemics in tropical gardens or fields.

The Indians who inhabit the Amazon basin cultivate large numbers of plants in their slash and burn agricultural systems. These systems have been studied frequently. Denevan and Treacy (1987), in a detailed description of a young slash and burn system in the Brazilian Amazon, identified 118 useful plants growing in the Bora Indians' fields. The Indians' primary staple was cassava, of which they have 22 cultivars, but they also cultivate maize, rice, sweet potatoes, cowpeas, peanuts, pineapple, plantain, peppers, yams, sugar cane, and cocoyams in their swidden fields.

Salik (1989) found 70 different species grown in yard gardens by the Amuesha Indians in the Peruvian upper Amazon. However, the number in any single garden varied, and the maximum number was 26 (Salik and Lundberg 1990). The number of different species planted depended on economic status, off-farm employment, and tenure and family status. For example, without land tenure, fruit trees were not planted, as other families might harvest the fruit.

Nations (1987), writing on Mayan Indian agriculture, noted: "Lacandon agroforestry combines up to 79 varieties of food and fiber crops on single-hectare garden plots cleared from forest." Alcorn (1984) found 300 species of useful plants in forest plots managed by Huastec Indians in Mexico. Mayan Indians in Mexico utilize a large number of plants, and have their own comprehensive plant classification system. Berlin et al. (1974), describing the Mayan (Tzeltal) taxonomic system, stated: "At this time, a total of 471 mutually exclusive generic taxa have been established as legitimate Tzeltal plant groupings."

Eden (1988) compared slash and burn systems in Colombia and Papua New Guinea. In Colombia, 11 swiddens were studied, in which 38 crop species were found. There was an average of 12 crops grown per swidden. Cassava constituted an average of 81% of the plants in a swidden. The level of diversity in the swiddens varied considerably. Greater levels of diversity were found in Papua New Guinea than in Colombia.

Christanty et al. (1986) observed that home gardens in Indonesia were highly diverse and contained a large number of species. They cited Hadikusumah as stating that he found 112 species in *kebun-talun* (a rotation system between a mixed garden and a tree planting) and 127 species in home gardens. The plants found included medicinal plants, ornamentals, spices, vegetables, fruits, firewood, building materials, and various cash crops.

In a survey in West Java Abdoellah and Marten (1986) found 235 useful species of plants grown in household gardens. The average garden had 20 species while the average upland field had eleven species. Michon et al. (1983) also studied the village gardens in West Java, and they observed that, although small in size (often less than 0.1 ha), over 70 plant species might be grown in the gardens. These included plants for food, timber, firewood, medicinal purposes, and ornamentals. The striking diversity of species used (some villages were reported to use up to 250 crop species) has important implications for the importance and severity of diseases in the gardens. Hawkes (1983) discussed kitchen gardens as sources of crop plant diversity, and suggested that the amount of variation in crop plants in such gardens still is close to what it might have been in ancient times. He suggested that this variation can still be found in household gardens in tropical areas.

As previously noted, Geertz (1963) suggested that swidden garden plots mimic the architecture of the tropical forest, especially in their high degree of diversity. A number of authors (Beckerman 1983a, 1983b, Hames 1983, Vickers 1983, Boster 1983, Stocks 1983) disagreed with some aspects of Geertz's model and reevaluated his concepts. They stated that the high degree of diversity suggested by Geertz is not found in most traditional systems studied. However, Weinstock (1984) pointed out that the studies of the above authors were primarily in South American forests where the primary crops are root and tuber crops such as cassava. In Asia, where Geertz made his studies, various grains such as rice are most important. Interspecific diversity would likely be more significant in influencing disease development in grain crops, as compared to asexually propagated crops such as cassava in which diversity is primarily intraspecific rather then interspecific.

Not only is there a considerable diversity of species in many traditional agroecosystems, but in some traditional systems even fields are dispersed, which provides an additional degree of protection from pests. Rhoades (1988) noted: "I have known farmers who have as many as 90 tiny fields scattered over a valley and frequently located several days' walk apart." These scattered fields would certainly provide some protection against plant disease epidemics. Further examples of the value of dispersion of host plants among other species to restrict the spread of pathogens are given in Chapter 19 on multiple cropping. Crop mixtures increase diversity and thus contribute to disease management (Burdon 1978).

Intraspecific Diversity

Clawson (1985) has discussed in detail intraspecific diversity among tropical crops. Many traditional farmers have maintained high levels of diversity within cultivars of a single crop species. Clawson called this "intraspecific polyculture" and gave examples of its occurrence with maize, cassava, yams, taro, beans, millet, rice, potatoes, and sweet potatoes.

In 1980 I travelled to Mexico with a group of students. On one visit we went to the farm of a traditional farmer near Puebla, Mexico. As the students were talking to him, I noticed a pot containing beans and separated out 17 different types from the pot (Figure 24.2). Later we found out that they included common beans (*Phaseolus vulgaris*), lima beans (*P. lunatus*), and scarlet runner beans (*P. coccineus*). The farmer said he grew all of them on his 1.5 ha farm. When asked why he grew so many varieties, he noted that some years it was wet, and some years it was dry. Some varieties did better in wet years, and some did better in dry years. Certain years, when insects attacked, some varieties survived while others did poorly. His wife preferred certain varieties for specific cooking purposes. The diversity of his many varieties doubtless gave him some protection against the vagaries of climate and biological sources of stresses. He mentioned nothing about disease reactions.

Watson (1983) described the large number of crop cultivars grown by Arabs in ancient times. He cited the ninth century writer al-Jahiz as stating that 380 kinds of dates were said to be found in the market of Basra, and also the tenth century writer Ibn Wahshiya as stating that the number of dates in Iraq could not be counted. According to al-Ansari, about 1400 A.D., near a small town on the North Africa coast there were 65 kinds of grapes, 36 kinds of pears, 28 kinds of figs, and 16 kinds of apricots. In ancient Egypt, the writer Abd al-Latif stated that the number of citrus varieties approached infinity. Although some exaggeration may have occurred, certainly, numerous cultivars of crop plants were common in Arab agriculture.

The work of Boster (1983, 1984a, 1984b, 1985, 1986) gives a fascinating example of intraspecific diversity. He has made detailed studies of the agriculture of the Aguaruna Jivaro Indians in the humid Amazon forests of Ecuador. Although they cultivate 50 other crops, their primary food is cassava (manioc), which comprises over 80% of the major crops grown. The Indians' swidden plantings exhibit much less interspecific diversity than do many other slash and burn systems (Boster 1983). The Aguaruna Jivaro men are hunters, while the women do the gardening. A collection of over 100 cultivars of cassava are maintained by the women. Boster (1984a) suggested that "The Aguaruna maintain diversity for its own sake without the need for clearly articulated pragmatic reasons for each variety maintained."

Boster did not discuss whether diseases occur on cassava. Nevertheless, diversity in respect to diseases and insects is doubtless important. He wrote: "They maintain a number of rarer varieties when cold logic might recommend abandonment in favor of higher yielding varieties. In maintaining diversity for its own sake, they show it does not pay to think too hard about which manioc varieties to maintain." In another paper, Boster (1985) presented evidence that the Aguaruna Jivaro have selected cultivars that can be distinguished on the basis of morphological characteristics. The yield of a particular cultivar is important to the Jivaro, but is not a major selection criterion. Other desired characteristics are good qualities for beer making, good storage and eating qualities, and rapid growth rate. If humans maintain diversity in their crop plants for diversity's sake, as the Jivaro do, and without thinking too hard about it, this will constitute a great contribution to future generations.

Barker (1984) described sweet potato production in the slash and burn system of the Ikalhans in upland Luzon in the Philippines. The Ikalhan group maintained 48 distinct types of sweet potatoes and 24 different types were sometimes planted in a single field along with other crops. Miracle (1967) described the primarily shifting agriculture of the many different tribal groups in the Congo basin of Africa in considerable detail. The Medje, for example, grew 80 varieties of the over 30 species of food crops they cultivated in 1911. They cultivated 27 varieties of bananas and plantains and 22 varieties of yams and related crops. The Azande grew 52 different crops, and the Banda had 60 cultivated plants.

A crop that has evolved considerable genetic diversity is rice. It may have first been cultivated by man as early as 8,000 B.C. in South and Southeast Asia, perhaps first in India (Chang 1976). Rice is primarily cultivated in the tropics and subtropics, but it is also important in temperate countries such as Italy, Japan, and Korea. During the thousands of years that man has cultivated rice, three main geographical "races" have evolved. These are *Indica,* grown in tropical areas, *Japonica,* the short grain rice grown in temperate areas, and *Javanica,* restricted to parts of Indonesia and the Philippines on rice terraces. Within these races there is tremendous diversity. Grist (1968) stated that there were thousands of cultivars grown in India. Indonesia has about 8,000 varieties (Bernsten et al. 1982), the Philippines 1,500, and Thailand 3,000 (Brush 1986). The International Rice Research Institute's rice collection in 1990 included about 85,000 accessions (IRRI 1990, Vaughan and Stich 1990). Dennis (1987) studied farmers' management of diversity in rice in Northern Thailand, and concluded that some farmers consciously maintained varietal diversity at levels above those needed for market and agronomic considerations.

Disease Resistance in Traditional Landraces

Buddenhagen (1983), Harlan (1976), and Leppik (1970) noted that the greatest number of genes for disease resistance are usually found in landraces, where the host and pathogen have coexisted for long periods of time. Zadoks and Schein (1979) wrote: "Whether disease resistance was positively selected by early farmers is conjectural, but certainly they selected for varieties with higher productivity." Obviously, if a disease killed a number of the plants that early farmers were cultivating, the survivors would probably be selected.

Most of the agriculture of Ethiopia can be characterized as "traditional." Harlan (1976) noted that although the dates that barley and wheat were introduced into Ethiopia are not known, apparently both species have been there long enough for both hosts and pathogens to have adjusted reasonably well to living together. Plant breeders found resistance in Ethiopian barley lines to powdery mildew (*Erysiphe graminis*), loose smut (*Ustilago segetum* var. *tritici*), leaf rust (*Puccinia hordei*), net blotch (*Drechslera teres*), Septoria (*Septoria nodorum*), scald (*Rhynchosporium secalis*), barley yellow dwarf virus, and barley stripe mosaic virus. When a large number of barley cultivars were tested for resistance to the barley yellow dwarf virus, 113 of those found to have resistance were from Ethiopia. Thus, Ethiopian farmers over the centuries selected landraces with resistance to many of the world's major barley diseases.

The studies and observations of Trutmann et al. (in press) in the highlands of Zaire near the border of Rwanda give a clear example of traditional farmer (all women) selection for diversity in beans (*Phaseolus vulgaris*) that provided a measure of disease resistance. The investigators hypothesized that if farmers selected for resistance to a particular disease, then the prevalence of resistant varieties in the traditional mixtures planted should vary in relation to the suitability of the environment for the development of the disease. They measured the amount of resistance to anthracnose (*Colletotrichum lindemuthianum*) in local farmers' mixtures, using farmers' bean seed collected from three different altitudes. Their studies showed that a rather large percentage of the varieties in the traditional mixtures had resistance to *C. lindemuthianum*. In addition, they found that the percentage of the traditional mixture that consisted of resistant varieties increased greatly with altitude and the prevalence of anthracnose. At an elevation of 1,500 m only 16% of the mixture consisted of resistant varieties, while at 2,000 m (where *C. lindemuthianum* is most damaging) 25% of the mixture was resistant. Thus, there were 51% more resistant genotypes in the high altitude mixture than in the lower altitude mixture. It is interesting that the figure of 25% is similar to data on crown rust of oats (*Puccinia coronata*) from Iowa, South Texas, and Israel showing that if about one third of an oat population is resistant to *P. coronata*, the population is protected from crown rust (personal communication -- A. J. Browning). Jeger et al. (1981) indicated that 25% resistance in mixtures reduced disease over the expected in wheat and barley disease systems. The excellent Trutmann et al. study (a cooperative effort between a plant pathologist and two anthropologists) clearly illustrated intraspecific selection for disease resistance by traditional farmers.

Mohamed and Teri (1989) described the mixtures of bean seeds of different cultivars made by traditional small-scale farmers in Mgeta, Tanzania. The mixtures contained seeds of different colors, shapes and sizes, and the composition of the mixture varied with location. A small number of farmers deliberately mixed seed of different cultivars. Allen et al. (1989) cited work by the ISNAR (International Service for National Agricultural Research) demonstrating that most exotic bean varieties were less adapted and more affected by diseases than were the mixture of local varieties used by traditional farmers in the Great Lakes Region of Africa. Allen et al. (1989) and van Rheneen (1979) found bean landraces of mixed seed to be common in Uganda, Tanzania, and Malawi.

Traditional rice farmers in many parts of Asia and Africa use mixtures of cultivars. In Madagascar rice mixtures have been planted for generations, and farmers believe the mixed populations give more stable yield (Bonman et al. 1986). Farmers in Indonesia, Bhutan, and Bangladesh were also reported to cultivate a mixture of genotypes. Experiments in the International Rice Research Institute in the Philippines with mixtures showed that rice blast (*Pyricularia oryzae*) caused less damage in mixed stands. Bonman et al. (1986) stated: "Leaf blast was significantly less in the mixed stands than in the pure stands of the most susceptible cultivar of each set, and was statistically equal to the amount of disease in the most resistant component in five of the six inoculated experiments."

Genetic Vulnerability

Mangelsdorf (1966) pointed out that humans have used about 3,000 plant species for food and that about 150 of these have entered into world commerce. He suggested that today humans are fed primarily by only about 15 species of plants. This dependency on such a small number of crops is a form of lessened genetic diversity. Crops grown in most developed countries are generally highly uniform, and much of their original genetic diversity has been lost.

Cowling (1978) listed 18 studies warning against the dangers of genetic uniformity in our major crops, but noted that these studies and their recommendations have not been adequately heeded. The genetic vulnerability of crops in the US (where relatively few varieties dominate the major crops) was illustrated by the Southern corn leaf blight epidemic (caused by *Cochliobolus heterostrophus*) of 1970 (Ullstrup 1972). That severe epidemic galvanized plant pathologists and breeders into a furious flurry of finger pointing at genetic vulnerability (National Academy of Sciences 1972). Since then, plant pathologists and plant breeders have made more serious attempts to incorporate diversity into modern breeding programs. Nevertheless, the crops in farmers' fields today in developed countries like the US are far from diverse, and most are highly uniform. The disappearance of maize germplasm in Latin America is discussed by Timothy (1972).

Following the Southern corn leaf blight epidemic, the National Academy of Sciences published *Genetic Vulnerability* (1972). Regarding rice, an article in the book noted:

> Large-scale crop failures due to disease and insect attacks have not occurred in the recorded history of the crop with the exception of the Bengal famine of 1943. One reason for this may be the multitude of varieties grown in all the major rice-growing areas of tropical Asia, a genetic diversity that may have served the useful purpose of suppressing the build-up of parasite and insect populations.

Perhaps as many as two million Indians died of starvation in the Bengal famine of 1943 (Padmanabhan 1973). The famine was attributed to the fungus *Cochliobolus miyabeanus* (causal agent of brown spot of rice), although some pathologists (National Academy of Sciences 1972) believe that by itself *C. miyabeanus* does little damage and that the disease is generally associated with other nutritional or physiological problems. This famine occurred before the Green Revolution in an area where there were a large number of different rice cultivars, so it probably does not constitute an example of genetic vulnerability. There were additional epidemics due to various rice pathogens before the advent of the high-yielding varieties in the 1960s (Barr et al. 1975), but whether these epidemics could be characterized as "large-scale" is perhaps debatable.

Nevertheless, there is concern that the widespread use of a few cultivars with common genes (e.g. for dwarfing in rice) may increase the chances for a new race of an existing pathogen or a now obscure disease or insect pest to cause widespread, serious losses in rice and wheat. The National Academy of Sciences report (1972) stated: "it appears unlikely that the common gene for dwarfing in all the new varieties would, *per se*, increase the genetic vulnerability to any disease or pest." Time will tell us whether this conclusion is correct. The Texas male sterile cytoplasm found in almost all US maize before the Southern corn leaf blight epidemic of 1970 was not thought to be susceptible to any pathogen either.

Teri and Mohamed (1988) stated: "widespread plant disease epidemics in traditional agriculture are either rare, undocumented, unnoticed or all three." They suggested that this is due in large part to the diversity of crops grown by traditional farmers.

In an interesting discussion of diversity and genetic vulnerability, Brown (1983) suggested that genetic diversity *per se* does not necessarily provide insurance for a species against genetic vulnerability. He used various examples to illustrate this point. First, the demise of the American chestnut by chestnut blight (*Cryphonectria parasitica*), which exterminated chestnuts from Eastern US in about two decades. The host species is highly variable genetically. Also, the highly variable American elm (*Ulmus americana*) has proven highly susceptible to *Ceratocystis ulmi* (Dutch elm wilt) as it has spread throughout the United States. Additionally, in the 1950s, southern rust of maize (*Puccinia polysora*) was introduced into East Africa where it spread rapidly on the open-pollinated maize varieties of the region. The local landraces of maize in East Africa were highly variable, but all were highly susceptible to *P. polysora*, and severe epidemics resulted. Brown gave still other examples, but made the important point that genetic diversity may be highly desirable, but unless the diversity includes genetic resistance to a specific organism it is of little value. Brown concluded his article by stating: "plant germplasm is among the most essential of the world's natural resources. Its conservation merits far greater attention than it is now receiving." Schmidt (1978) also noted that diversity is no safeguard against pathogens, and also used the chestnut blight epidemic as an example. The point should also be made that uniformity is not necessarily dangerous unless it affects resistance.

Genetic Erosion

Natural plant communities rarely have the serious plant disease epidemics that frequently occur in the intensive and extensive monocultures of our modern agriculture. Erosion of genetic resistance to pathogens and pests, due to a few new varieties of crop plants replacing landraces, is a major concern today in the international agriculture community. I have certainly done my part to contribute to genetic erosion, as I worked for eleven years in Colombia with the Rockefeller Foundation in cooperation with the ICA (Instituto Colombiano Agropecuario -- Colombian Agricultural Institute) potato and plant pathology research programs. This and subsequent ICA programs introduced many new and successful potato varieties in Colombia. Gomez and van de Zaag (1986) stated: "Currently, less than 15% of the potato area in Colombia is cropped with old cultivars, and these persist mainly because of their outstanding cooking quality and taste." Although the new varieties are often much higher yielding, and may have other desirable characteristics, this rapid change to newer potato varieties threatens the survival of varieties selected for centuries by indigenous populations. A similar scenario could be given for several of the worlds' major food crops, such as wheat, rice, and maize. Thus, many of the invaluable genetic resources embodied in landraces that were developed in a long evolutionary process may eventually disappear (Thurston 1980).

Termed the "Green Revolution", the rapid spread and use of high-yielding varieties (HYVs) of wheat and rice has been spectacular. For example, over 75% of the rice area in the Philippines is planted to HYVs developed since 1965 (Herdt and Capule 1983). From 49,000 ha in the 1965/1966 crop year, the area planted to the HYVs of wheat and rice in Asia and the Near East had by 1976/77 increased to almost 55 million ha (Dalrymple 1978). In 1980, almost 40% of the rice produced in South and Southeast Asia consisted of HYVs, and they were estimated to contribute 4.5 billion dollars annually to the value of rice produced in Asia (Herdt and Capule 1983). The rapid adoption of the the high-yielding varieties of wheat and rice is one of the major success stories in the history of agriculture; however, there is some concern in the international agriculture community that much valuable diversity is being lost as traditional cultivars are replaced and discarded.

Limitations of Modern Plant Breeding

In spite of what plant breeders and those involved in "genetic engineering" might imply about the immense power of their disciplines to improve existing crops, there are instances when breeders have not been able to improve on cultivars selected by traditional farmers centuries ago. One of the most clearcut cases of such failure is found with pineapples. The Pineapple Breeding Institute of Hawaii, probably the major pineapple breeding program in the world, produced and tested millions of seedlings in one of the most modern plant breeding efforts of the time in an attempt to improve on the standard cultivar "Cayenne" (Buddenhagen 1977). Cayenne was selected in Venezuela by the Maipure Indians before Columbus discovered the Americas (Purseglove 1972). Pineapple production today is still based on the asexually propagated progeny of five plants taken from Cayenne, French Guiana, to France in 1820. Cayenne is still used all over the world by the industry today, and Hawaii's pineapple breeding program has been largely abandoned. According to Buddenhagen (1977) varieties with superior horticultural characteristics selected from the breeding program were more susceptible to diseases than Cayenne, and therefore not released.

Strategies for reducing dependence on specific resistance (one or a few genes), such as the use of general or horizontal (multigenic) resistance or the use of multiline varieties (Borlaug 1958, Browning 1975, Browning and Frey 1969, Browning and Frey 1981, Jensen 1952), have been proposed. Discussions of these strategies for avoiding genetic vulnerability in agriculture are given by Buddenhagen 1983, Cowling 1978, Day 1977, Fry 1982, Jenkyn and Plumb 1981, National Academy Sciences 1972, Nelson 1973, and Vanderplank 1963, 1968, 1975.

Conservation of Genetic Resources

How best to conserve the invaluable genetic resources found in traditional agricultural systems is the subject of much talk and little action. Conservation of crop germplasm is only part of a worldwide problem of maintaining our disappearing genetic diversity in many wild plant and animal species (Timothy 1972, Wilson 1988a). The facilities for conservation of crop germplasm in the US and other developed countries are generally poorly funded, considering the magnitude and the urgency of the problem. For example, the International Board for Plant Genetic Resources (IBPGR) of the CGIAR (Consultative Group in International Agricultural Research) has its headquarters in Rome, Italy, and assists in the collection, conservation, documentation, and exchange of crop genetic material (Plucknett et al. 1983, Williams 1988). Core operating expenditures of the IBPGR in 1987 were only 5.1 million dollars (CGIAR 1988). This is a miserable sum when one considers the importance and magnitude of what needs to be done!

The various international centers concerned with crops collect and maintain germplasm collections. For example, the International Potato Center (CIP) in Peru has one of the largest collections of potato (Solanum spp.) germplasm in the world and is still collecting throughout the Americas so as to add to their collection before valuable germplasm is lost. Probably only 5% of the genetic variability of Solanum is found in the cultivars now in use throughout the world (CGIAR 1974). Future generations may find that the preservation of potato germplasm alone more than justified all the financial support given to the CIP since it was founded (Thurston 1980). Today, I would add that the preservation of potato germplasm may justify all of the financial aid given to the entire CGIAR system.

The conservation of landraces within existing traditional agricultural systems is a controversial subject (Altieri and Merrick 1988, Browning 1975, Oldfield 1984, Oldfield and Alcorn 1987, Williams 1983). Browning (1975) recommended the establishment of "living-gene parks", to maintain not only diversity in crops, but also to maintain diversity in their pathogens. In indigenous ecosystems and traditional agriculture systems germplasm can be maintained and collected, knowledge on how plants protect themselves can be obtained, and the continuing process of coevolution can be studied. Many scientists and various groups have suggested *in situ* conservation as the ideal, but some suggest that such a solution would be impossible and/or impractical, as, according to Frankel (1970), "One may be doubtful whether such attempts are likely to be successful or, in view of their cost, even worthwhile." Considering the potential importance of such efforts to future plant improvement, this seems to be an overly negative attitude. Oldfield and Alcorn (1987) suggested that the world's genetic resources might be maintained by using three approaches: 1) *in situ* conservation combined with sound conservation-development; 2) *in situ* conservation combined with research centers; and 3) *ex situ* reserves. An example of an *in situ* conservation-development approach is the Kuna Yala Indigenous Reserve in Panama, which includes 60,000 ha of tropical forests (Oldfield and Alcorn 1987). Panamanian forests are highly diverse, and Raven

(1981) related that there are about as many plants in Panama as in all of Europe. The Kuna Indians are receiving advice from organizations such as UNESCO, the Smithsonian Tropical Research Institute, and the Centro Agronomico Tropical de Investigacion y Enseñanza (CATIE) on the development in the reserve, which includes both wilderness conservation and rural development. There are also plans to preserve the ethnobotanical and agricultural knowledge of the Kuna Indians in this effort. Obviously, there are many difficult problems with such approaches, but the effort is certainly worth the concern of our best scientists, as time is running out for humans to preserve a heritage of 10,000 years of crop selection.

Summary

Many of the major crops on which humans depend for food are constituted primarily of cultivars or landraces selected long before "modern" agricultural science began. These landraces are usually genetically diverse and in balance with the environment and endemic pathogens. They are dependable and stable in that, although not necessarily high yielding, they yield some harvest under all but the worst conditions. Numerous centers of diversity for major food crops are still found, primarily in developing countries. Cultivar selection and maintainence by traditional farmers still continues in some indigenous societies. A major question is, how can the yields of traditional farmers be increased without destroying valuable traits like stability and resistance to various pests.

An interesting study and analysis of natural ecosystems and agroecosystems relative to plant diseases was made by Browning (1981). He suggested: "Emulating nature's model can improve the effectiveness of crop protection systems with minimum loss of yield and quality traits. Thus epidemics do not usually occur if an agroecosystem has minimum diversity so that the pathogen retains some of its natural self-regulating ability. " Compared to most traditional systems, there is little species diversity in most modern farming systems. Agricultural scientists must balance the advantages of high levels of interspecific and intraspecific diversity for managing plant diseases and maintaining yield stability provided by many traditional farming systems against the positive economic advantages but serious risks of modern agricultural practices and the new high-yielding varieties. A challenge to future plant pathologists will be to develop plant disease management strategies that incorporate traditional levels of diversity in a sustainable agroecosystem.

In the long run, the CGIAR (Consultative Group in International Agricultural Research) centers may make one of their major contributions by collecting and safeguarding diversity in the germplasm of the world's major food crops. The survival of the diversity incorporated in the many traditional landraces and their wild relatives, most frequently found in developing countries, is seriously threatened by numerous human activities. The loss of these irreplaceable reservoirs of genetic material would constitute a tragedy for humans. There are hopeful signs that the value of such germplasm is coming to be appreciated, but financial support for its preservation is still meager.

The genetic erosion caused by the rapid spread and spectacular success of the new, high-yielding varieties (HYVs), especially of wheat and rice, has caused considerable concern in the international agricultural community. Much valuable diversity in several major crop species is being lost as landraces are being supplanted by a few "improved" varieties. The widespread use of a few cultivars with common genes also increases the possibility of serious epidemics. To date, no widespread, disastrous epidemics have resulted from the introduction of the HYVs of rice or wheat; however, localized epidemics on both wheat and rice have occurred. The CGIAR (Consultative Group in International Agricultural Research) network of institutes and other international agencies carefully monitor the disease situation for these crops and are prepared to respond to evidence of a breakdown of resistance (Saari and Wilcoxson 1974).

Considering the importance of conserving traditional crop germplasm and the magnitude of the task, the efforts to date in this regard have been sadly inadequate. Although the CGIAR-funded international centers are making efforts to conserve germplasm of a few major crops, their efforts are only a small part of a worldwide problem of maintaining diversity in many crop and wild species of plants.

25

Recommendations

The following recommendations are a synthesis of practices that might be utilized by both modern and traditional farmers for the management of plant diseases.

1. Historically, many sustainable agricultural systems incorporated large quantities of organic matter into soil. This incorporation generally resulted in less soilborne disease, in addition to other important agronomic benefits, and the practice should be recommended whenever feasible. (Chapters 3, 12, 14, 15, 16, 17, 20)

2. Some diseases are suppressed by shade, whereas others increase in importance under shade. Manipulation of shade should be considered as a possible component of disease management systems. (Chapter 22)

3. The use of antagonistic plants (trap crops and trap plants) should be considered for the management of nematodes and other soilborne pathogens. (Chapters 3 and 13)

4. Clean seed or healthy propagating material, or such material treated to kill pathogens, often has positive and dramatic effects on plant health and crop yields. Heat and anaerobic cold water treatments can sometimes kill seedborne pathogens. Healthy planting materials should always be utilized when feasible. (Chapter 8)

5. Plant pathogens are often transmitted when vegetative propagating material is cut. Using sterile tools for cutting propagating material or the use of uncut propagation materials are important practices. For example, planting whole rather that cut potato tubers prevents losses due to fungi and bacteria that occur when tubers are cut. (Chapter 21)

6. The use of seed beds, and the subsequent transplanting of carefully selected healthy seedlings, is a useful practice for many crops. (Chapters 8 and 14)

7. The density of crop or plant stands has important effects on disease incidence and severity. Dense plant stands generally increase disease, but in some cases (i.e. with some virus diseases) may reduce disease. Crop density can be altered by manipulations of plant and row spacing (i.e. the rate of sowing and planting). (Chapters 4 and 19)

Crop density and architecture can also be altered by practices such as pruning, thinning, trellising, fertilization, water management, staking, and harvesting plants or plant parts. Avoiding foliage or root contact can also reduce the incidence of some diseases. (Chapter 4)

8. The depth at which seeds and propagating materials are planted may affect disease incidence or severity and should be looked into when designing disease management strategies. Shallow planting is often an effective disease management practice, as plants emerge from the soil quickly when not planted too deeply. Plants are especially susceptible to disease during germination and emergence of seedlings from the soil. (Chapter 5)

9. Fallow periods are often beneficial in reducing losses from plant diseases, especially soilborne diseases. For disease management, fallowing is generally more effective in combination with rotations. Both dry and flood fallowing should be considered in planning plant disease management. (Chapters 9 and 11)

10. Fire and heat are often overlooked as plant disease management practices. The high temperatures produced by burning may eliminate inoculum of many pathogens. The negative effects of burning should not be overlooked when contemplating the use of burning for plant disease management. (Chapter 10)

11. Traditional agriculture has utilized flooding extensively for the management of plant pathogens. For example, the paddy rice system, in addition to its various agronomic benefits, has an important role in reducing the importance of soilborne diseases. Where feasible, flooding should be recommended as a possible plant disease management practice. (Chapter 11)

12. The many ingenious and effective practices for crop harvesting and storage developed by traditional farmers over the centuries warrant careful study and evaluation, as some of their practices might be more practical and appropriate for traditional conditions than the storage practices current in modern agriculture. (Chapter 18)

13. Mulches reduce plant diseases by reducing soil splashing, influencing the moisture content and temperature of the soil, and enhancing the microbiological activities that suppress plant pathogens. In the hot humid tropics, where plant growth is rapid and luxurious, the use of green manures and natural vegetation as mulches (as in the slash/mulch system) should be considered, as their use provides an effective management practice for some diseases. (Chapter 12)

14. It is difficult to generalize with any degree of accuracy about disease management through the use of multiple cropping. Recommendations on multiple cropping should be thoroughly tested, and site-specific recommendations will often be necessary. Time tested local practices should serve as the first guide to recommendations. Nevertheless, most of the literature indicates that there is less disease in most types of crop associations than in monoculture. (Chapter 19)

15. Multistory systems often existed for centuries in tropical areas without major disease problems. Combining manipulations of plant architecture and shade, the use of landraces, and a diversity of species, multistory systems may provide useful models for other areas in the tropics. (Chapter 20)

16. Raised fields, raised beds, ridges, and mounds were used widely by traditional farmers for millennia. Better drainage and irrigation, enhanced fertility, and frost control are other important benefits of these systems, but planting in soil raised above the soil surface is also an important disease management practice for soilborne pathogens. (Chapter 14)

17. The use of rotation should be carefully investigated and utilized in schemes designed to aid traditional farmers, keeping in mind that the value of crop rotation for the management of specific diseases is highly location specific. In addition to agronomic benefits, such as improved soil texture and better use of nutrients and water, rotations contribute to the management of soilborne pathogens. Rotations are probably the best documented traditional disease management practice. (Chapter 15)

18. Sanitation practices include removing, roguing, or destroying plants or plant parts that may be sources of initial or secondary inoculum. Flooding, burning, composting, many tillage practices, and production of clean seed are other examples of sanitation practices that result in enhanced disease management. Some effective labor-intensive sanitation practices may be feasible for traditional farmers, whereas labor costs in modern agriculture might make such practices prohibitively expensive. (Chapter 21)

19. Proper selection of planting dates is of great importance in the management of many plant diseases, and, although little information is available on the selection of planting dates by traditional farmers to reduce disease incidence, the practice should be carefully considered in all disease management schemes. (Chapter 6)

20. Site or habitat selection is important in the management of plant diseases. Different regions or altitudes can often be selected for crops, so that disease severity is reduced because of climatic conditions unfavorable to pathogens. Within the farm, sites may be chosen or avoided because of previous crops (rotation), soil type, air or water drainage conditions, or previous history of disease. (Chapter 7)

21. Minimum tillage may increase, decrease, or have no effect on plant diseases. Some pathogens survive in crop residues in minimum tillage systems, whereas others are inhibited in such systems. Because of their additional agronomic benefits, such systems should be incorporated for disease management whenever feasible. (Chapter 17)

22. Tillage practices have various effects on the incidence and severity of disease. For example, deep plowing may bury plant pathogens from the topsoil into deeper layers of the soil where they cause less or no disease. Organic matter may be incorporated into the soil by tillage. Successive tillage operations can reduce the inoculum of some pathogens by solarization. Tillage practices should be investigated for their possible value in disease management. (Chapter 17)

23. The use of drying agents (ashes, chalk) for crop storage and natural or non-toxic pesticides for control of insect vectors and pathogens are effective techniques of traditional farmers. Their use should be encouraged wherever feasible as alternatives to toxic pesticides. (Chapter 2)

24. Weeds may serve as a reservoir of inoculum for various diseases and a means for pathogen inoculum to survive adverse conditions and overseason. Weed therefore often play an important role in disease epidemiology. Weed growth may increase humidity and shade within a crop and thereby affect disease incidence. The effect of weeds on plant disease management should always be considered. (Chapter 23)

25. The high levels of interspecific and intraspecific diversity found in many traditional farming systems have well-proven advantages for managing plant diseases and maintaining yield stability. The advantages of this diversity must be balanced against the positive economic advantages but serious risks (genetic vulnerability to diseases and other pests) of using the new high-yielding varieties. (Chapter 24)

26. Many of the major crops on which humans depend for food are constituted primarily of cultivars or landraces selected before "modern" agricultural science began. These landraces are usually genetically diverse and in balance with the environment and endemic pathogens. They are dependable and stable in that, although not necessarily high-yielding, they yield something under all but the worst conditions. The conservation and possible utilization of these landraces should be considered a priority in disease management schemes. (Chapter 24)

Bibliography

Abawi, G. 1989. Root rots. pp. 105-157. In: Schwartz, H. F. and M. A. Pastor-Corrales, ed., *Bean production problems in the tropics.* CIAT, Cali, Colombia.

Abawi, G. S., D. C. Crosier, and A. C. Cobb. 1985. Root rot of snap beans in New York. N. Y. Food and Life Sci. Bull. 110, Cornell Univ., Ithaca, NY. 7 pp.

Abawi, G. S. and M. A. Pastor-Corrales. 1990. *Root rot of beans in Latin American and Africa: Diagnosis, research methodologies, and management strategies.* CIAT, Cali, Colombia. 114 pp.

A'Brook, J. 1964. The effect of planting date and spacing on the incidence of groundnut rosette disease and the vector, Aphis craccivora Koch., at Mokwa, Northern Nigeria. Ann. Appl. Biol. 54:199-208.

______. 1968. The effects of spacing on the number of aphids trapped over the groundnut crop. Ann. Appl. Biol. 61:289-294.

Abdoellah, O. S. and G. G. Marten. 1986. The complementary roles of homegardens, upland fields, and rice fields for meeting nutritional needs in West Java. pp. 293-325. In: Marten, G. G. ed., *Traditional Agriculture in Southeast Asia. A Human Ecology Perspective.* Westview, Boulder, Co.

Adams, R. E. W., W. E. Brown Jr., and T. P. Culbert. 1981. Radar mapping, archaeology, and ancient Maya land use. Science 213:1457-1463.

Adegbola, M. O. K. 1977. The reactions of four Nigerian isolates of cacao swollen shoot virus (CSSV) on the germination and growth of cacao seedlings under different environmental conditions. Proc. V Intl. Cocoa Res. Conf., Ibadan, Nigeria. pp. 338-343.

Adesiyan, S. O. and M. O. Adeniji. 1976. Studies on some aspects of yam nematode (Scutellonema bradys). Ghana J. Agr. Sci. 9:131-136.

Agarwal, V. K. and J. B. Sinclair. 1987. *Principles of Seed Pathology.* 2 Vol. CRC Press, Boca Raton, FL. 344 pp.

Ahenkorah, Y., G. S. Akrofi and A. K. Adri. 1974. The end of the first cocoa shade and manurial experiment at the Cocoa Institute of Ghana. J. Hort. Sci. 49:43-51.

Ahlgren, I. F. 1974. The effects of fire on soil organisms. pp. 47-72. In: Kozlowski, T. T. and C. E. Ahlgren. *Fire and Ecosystems.* Academic Press, New York.

Ainsworth, G. C. 1981. *Introduction to the History of Plant Pathology .* Cambridge Univ. Press, Cambridge. 315 pp.

Aiyer, A. K. Y. N. 1949. Mixed cropping in India. Indian J. Agric. Sci. 19(4): 439-543.

Akobundo , I. O. ed. 1980. *Weeds and their control in the humid and subhumid tropics.* IITA, Ibadan, Nigeria. 421 pp.

______. 1984. Advances in live mulch crop production in the tropics. In: Proc. Western Soc. Weed Sci. 37: 51-57.

Akobundu, I. O. and A. E. Deutsch, ed., 1983. *No-tillage Crop Production in the Tropics*. Proc. Symp. Monrovia, Liberia. August 1981. Intl. Plant Protec. Center, Oregon State Univ., OR. 235 pp.

Al-Awam, Ibn (Abu Zacaria Iahia). 1988. *Libro de Agricultura.* 2 Vols. J. A. Banqueri. (trans.) Clasicos Agrarios, Min. Agric., Pesca y Alimentación, Madrid. 1454 pp.

Alcorn, J. B. 1984. Development policy, forests, and peasant farms: reflections on Huastec-managed forests' contributions to commercial production and resource allocation. Econ. Bot. 38:389-406.

Alexander, M. 1977. *Introduction to Soil Microbiology.* Wiley, New York. 476 pp.

Allen, D. J. 1977. Intercropping and disease. pp. 34-35. In: IITA. Ann. Report 1976. IITA, Ibadan.

______. 1983. *The Pathology of Tropical Food Legumes. Disease Resistance in Crop Improvement.* Wiley, New York. 413 pp.

______. 1989. The Influence of Intercropping With Cereals on Disease Development in Legumes. CIMMYT/CIAT Workshop on Res. Methods for Cereal/Legume Intercropping in Eastern and Southern Africa. CIAT Regional Bean Program, Arusha, Tanzania. pp. 18.

Allen, D. J. and R. A. Skipp. 1982. Maize pollen alters the reaction of cowpea to pathogens. Field Crops Res. 5:265-269.

Allen, L.H., T. R. Sinclair, and E. R. Lemon. 1976. Radiation and microclimate relationships in multiple cropping systems. pp. 171-200. In: Papendick, R. I., P. A. Sanchez and G. B. Triplett, ed., *Multiple Cropping.* ASA Special Publ. 27, Am. Soc. Agron., Madison, WI.

Allen, D. J., M. Dessert, P. Trutmann and J. Voss. 1989. Common beans in Africa and their constraints. pp. 9-31. In: Schwartz, H. F. and M. A. Pastor-Corrales. *Bean production problems in the Tropics.* CIAT, Cali, Colombia.

Allen, W. 1965. *The African Husbandman.* Oliver and Boyd, London. 505 pp.

Alonso de Herrera, G. 1988. *Agricultura General.* Terrón, E., ed., Min. Agric. Pesca y Alimentación, Madrid. 445 pp.

Almazan, A. M. and R. L. Theberge. 1989. Influence of cassava mosaic virus on cassava leaf-vegetable quality. Trop. Agric. (Trinidad): 66: 305-308.

Altieri, M. A. and M. Liebman. 1986. Insect, weed, and plant disease management in multiple cropping systems. pp. 183-218. In: Francis, C. A. ed., *Multiple Cropping Systems.* MacMillan, New York.

Altieri, M. A. and L. C. Merrick. 1987. In situ conservation of crop genetic resources through maintenance of traditional farming systems. Econ. Bot. 41:86-96

Altieri, M. A. and L. C. Merrick. 1988. Agroecology and in situ conservation of native crop diversity in the third world. pp. 361-369. In: E. O. Wilson. ed., *Biodiversity*. Nat. Acad. Press.

Alvim de T., P. 1977. Cacao III. Climate. pp. 279-313. In: Alvim de T., P. and T. T. Kozlowski, eds., *Ecophysiology of Tropical Crops.* Academic Press, New York.

American Society Agronomy. 1978. *Crop Residue Management Systems.* ASA Special Publ. No. 31. Am. Soc. Agron., Madison, WI. 248 pp.

Amin, K.S. and J. C. Katyal. 1979. Incidence of leaf blast in relation to seedling density and chemical composition of rice varieties. Phytopath. Zeitschrift. 96:65-70.

Andrews, F. W. 1937. Investigations on black-arm disease of cotton under field conditions II. The effect of flooding infective cotton debris. Empire J. Exp. Agric. 5:204-218.

Anon. 1974. *Multiple Cropping Systems in Taiwan.* Food and Fertilizer Technol. Center for the Asian and Pacific Region. Taipei, Taiwan. 77 pp.

Antón Ramirez, D. B. 1865. *Diccionario de Bibliografía Agronómica.* Ministerio de Agric., Pesca y Alimentación, Madrid. 1015 pp.

Antonovics, J. and D. A. Levin. 1982. The ecological and genetic consequences of density-dependent regulation in plants. Ann. Rev. Ecol. Syst. 11:411-452

Araya V., R. and W. Gonzalez. 1987. *El Frijol Bajo el Sistema Tapado en Costa Rica.* Ciudad Univ. Rodrigo Facio, San José, Costa Rica. 272 pp.

Arevalo-R., J. and J. J. Jimenez O. 1988. Nescafe (Stizolobium pruriens (L.) Medic. var. utilis Wall ex Wight) como un ejemplo de experimentación campesina en el tropico humedo Mexicano. pp. 75-89. In: Del Amo R., Silvia. *Cuatro Estudios Sobre Sistemas Tradicionales.* Instituto Nacional Indigenista, Mexico, D.F., Mexico.

Armillas, P. 1971. Gardens on swamps. Science 174:653-661.

Arneson, P. A. 1971. Guia para la producción de maní (cacahuate) en Nicaragua. Dept. de Invest. Trop., United Fruit Co., San Pedro Sula, Honduras.

Asare-Nyako, A. 1977. Phytophthora palmivora. pp. 83-86. In: Kranz, J., H. Schmutterer and W. Koch, ed., *Diseases, Pests, and Weeds in Tropical Crops.* Verlag Paul Parey, Berlin.

Asher, M. J. C. and P. J. Shipton, 1981. Biology and control of take-all. Academic Press, New York. 538 pp.

Asomaning, E. J. A. and R. G. Lockard. 1964. Studies on the physiology of cocoa (Theobroma cacao L.). I. Suppression of swollen shoot virus symptoms by light. Ann. Appl. Biol. 54:193-198.

Autrique, A. and M. J. Potts. 1987. The influence of mixed cropping on the control of potato bacterial wilt. Ann. Appl. Biol. 111:125-133.

Ayala, A. 1968. Nematode problems of pineapple in Puerto Rico. pp. 49-60. In: Smart Jr., G. C. and V. G. Perry eds., *Tropical Nematology.* Univ. of Florida Press, Gainesville, FL.

Ayres, P. G. and L. Boddy. 1986. *Water, Fungi and Plants.* Cambridge Univ. Press,Cambridge. 413 pp.

Baker, K. F. 1938. Effect of raised beds in the reduction of heart rot and root rot in H. P. 4718. Pineapple Res. Inst. Rep. Pathol. Project 14f. Pineapple Res. Inst., Hawaii. 1 p.

Baker, K. F. and R. J. Cook. 1974. *Biological Control of Plant Pathogens.* Freeman, San Francisco. 433 pp.

Bandy, Dale E. and P. A. Sanchez. 1986. Post-clearing soil management alternatives for sustained production in the Amazon. pp. 347-361. In: Lal, R., P.A. Sanchez, and R. W. Cummings Jr. 1986. *Land Clearing and Development in the Tropics.* Balkema, Rotterdam.

Barker, T. C. 1983. PAFID (Philippine Assoc. for Intercultural Develop.) Field

Notes 2:6-7.

______. 1984a. Soil conservation among the Ikalahan. PAFID Fieldnotes 3:4-6. Philippine Assoc. for Intercultural Develop. Quezon City, Philippines.

______. 1984b. Shifting cultivation practices among the Ikalahans. UPLB-PESAM Working Ser. No. 1. Program on Environ. Sci. and Manage., Univ. Philippines at Los Baños, College, Laguna. pp. 1-30.

______. 1990. Agroforestry in the tropical highlands. pp. 195-227. In: MackDicken, K. G. and N. T. Vergarra. *Agroforestry: Classification and Management*. Wiley, New York.

Barnett, M. L. 1969. Subsistence and transition in agricultural development among the Ibaloi in the Philippines. pp. 284-295. In: Wharton Jr., C. R. ed., *Subsistence Agriculture and Economic Development*. Aladine, Chicago.

Barrau, J. 1958. *Subsistence Agriculture in Melanesia.* Bishop Museum Bull. 219. Honolulu. 111 pp.

Barr, B. A., C. S. Koehler and R. F. Smith. 1975. *Crop losses: Rice; Field Losses to Insects, Diseases, Weeds and Other Pests.* Univ. Calif. Berkeley/U.S. Agency Int. Dev., Pest Management and Related Environmental Protection Proj. Rep. 64 pp.

Barrera, A., A. Gómez-Pompa and C. Vázquez-Yanes. 1977. El manejo de las selvas por los Mayas. Biotica (Mexico) 2(2):47-61.

Barrerio, J. ed. 1989. *Indian Corn of the Americas. Gift to the World.* Northeast Indian Quarterly, American Indian Program, Cornell Univ., Ithaca, NY. 96 pp.

Barros, N. O. 1966. Valor de pas practicas culturales como metodo para reducir la incidencia de Monilia en plantaciones de cacao. Agric. Trop. (Colombia) 22: 605-612.

Barta, A. L. and A. F. Schmitthenner. 1986. Interaction between flooding stress and Phytophthora root rot. Plant Dis. 70:310-313.

Bartlett, H. H. 1955. *Fire in Relation to Primitive Agriculture and Grazing in the Tropics; an Annotated Bibliography.* Dept. of Bot., Univ. of Michigan, Ann Arbor. Part I (1955). 568 pp. Part II (1957). 873 pp.

______. 1956. Fire, primitive agriculture, and grazing in the tropics. pp. 692-714 In: Thomas Jr., W. L. et al eds., *Man's Role in Changing the Face of the Earth.* Univ. Chicago Press, Chicago.

Bassal, Ibn. 1955. *Libro de Agricultura.* J. M. Millas Vallicrosa and Mohamed Aziman (eds. and trans.) Instituto Muley El-Hasan, Tetuán. 231 pp.

Bean, G. A. and R. Echandi. 1989. Maize mycotoxins in Latin America. Plant Dis. 73:597-600.

Beaver, P. C., R. C. Jung and E. W. Cupp. 1984. *Clinical Parasitology*. 9th Ed. Lea and Febiger, Philadelphia. 825 pp.

Beckerman, S. 1983a. Does the swidden ape the jungle? 1983. Human Ecol. 11: 1-12

______. 1983b. Bari swidden gardens: crop segregation patterns. Human Ecol. 11: 85-101.

______. 1987. Swidden in Amazonia and the Amazon rim. pp. 55-94. In: Turner, B. L. II and Stephen B. Brush. *Comparative Farming Systems.* Guilford Press, New York.

Beer, J. 1987. Advantages, disadvantages and desirable characteristics of shade trees for coffee, cacao and tea. Agroforestry Systems 5(1):3-14.

______. 1988. Litter production and nutrient cycling in coffee (Coffea arabica) or cacao (Theobroma cacao) plantations with shade trees. Agroforestry Systems 7: 103-114.

Beets, W.C. 1982. *Multiple Cropping and Tropical Farming Systems.* Westview Press, Boulder, CO, 156 pp.

Belcher, J. V. and R. S. Hussey. 1977. Influence of Tagetes patula and Arachis hypogea on Meloidogyne incognita. Plant Dis. Reptr. 61: 525-528.

Bellod, M. 1947. La yesca de la vid en la región de Levante, su influencia en la longevidad de las cepas y experiencias sobre su tratamiento. Bol. Patología Vegetal y Entom. Agricola 15: 223-252.

Bentley, J. W. 1989. What farmers don't know can't help them: The strengths and weaknesses of indigenous technical knowledge in Honduras. Agric. Human Values 6: 25-31.

Berger, R. D. 1975. Disease incidence and infection rates of Cercospora apii in plant spacing plots. Phytopathology 65:485-487.

Berlin, B. D. E. Breedlove, and P.H. Raven. 1974. *Principles of Tzeltal Plant Classification. An Introduction to the Botanical Ethnography of a Mayan-speaking People of Highland Chiapas.* Academic Press, New York. 660 pp.

Bernsten, R. H., B. H. Siwi, and H. M. Beachell. 1982. The development and diffusion of rice varieties in Indonesia. IRRI Res. Paper Ser. No. 71. IRRI, Los Baños, Philippines.

Besse, J. 1972. Comparaison de deux méthodes d'etalbissment de Cacaoyéres (Comparison of two methods of establishing cacao). Cafe, Cacao, Thé 16: 317-332.

Beven, K. and P. Carling. 1989. *Floods. Hydrological, Sedimentological and Geomorphological Implications.* Wiley, New York. 290 pp.

Bigger, M. 1981. Observations on the insect fauna of shaded and unshaded Amelonado cocoa. Bull. Entomological. Res. 71: 107-119.

Birchfield, W. and F. Bistline. 1956. Cover crops in relation to the burrowing nematode, Radopholus similis. Plant Disease Reporter 40: 398-399

Bird, A. F. 1978. Root-knot nematodes in Australia. CSIRO Aust. Div. Hort. Res. Tech. Paper No. 2.

Blanco Galdos, O. 1981. Recursos genéticos y technologia de los Andes Altos. pp. 297-303. In: Novoa B., A. R. and J. L. Posner, ed., *Seminario Internacional Sobre Produccion Agropecuaria y Forestal en Zonas de Ladera de America Tropical.* Informe Technico No. 11. CATIE, Turrialba, Costa Rica.

Blanco Lopez, M. A., R. M. Jimenez Diaz and J. M. Caballero. 1984. Symptomology, incidence and distribution of Verticillium wilt of olive trees in Andalucia. Phytopath. Medit. 23:1-8.

Bock, K. R. and E. J. Guthrie. 1977. Cassava mosaic. pp. 10-13. In: Kranz, J., H. Schmutterer, and W. Koch, ed., *Diseases, Pests and Weeds in Tropical Crops.* Verlag Paul Parey, Berlin.

Bolens, Lucie. 1972. Engrais et protection de la fertilité dan l'agronomie hispano-arabe XI-XII siécles. Etudes Rurales 46: 34-60.

Bonman, J. M., B. A. Estrada and R. I. Denton. 1986. Blast management with upland rice cultivar mixtures. pp. 375-382. In: IRRI. *Progress in Upland Rice Research.* IRRI, Los Baños, Philippines.

Bonman, J. M., L. M. Sanchez and A. O. Mackill. 1988. Effects of water deficit on rice blast II. Disease development in the field. J. Plant Protec. Tropics 5(2): 67-73.

Boosalis, M. G., B. L. Doupnik, and J. E. Watkins. 1986 . Effect of surface tillage on plant diseases. pp. 389-408. In: Sprauge, M. A. and G. L. Triplett eds., *No-tillage and Surface-tillage Agriculture. The Tillage Revolution.* Wiley, New York.

Booth, R. H. 1977. A review of root rot diseases in cassava. In: Brekelbaum, T., A. Bellotti and C. J. Lozano eds., *Cassava Protection Workshop.* Cali, Colombia. Proc. CIAT, Cali, Colombia.

Booth, R. H. and D. G. Coursey. 1972. Storage of tropical horticultural crops. Span 15 (3): 1-3.

Borlaug, N. E. 1958. The use of multilineal or composite varieties to control airborne epidemic disease of self-pollinated crop plants. pp. 12-26. In: Proc. First Intl. Wheat Genetics Symp., Winnipeg.

Bornemisza, E. 1982. Nitrogen cycling in coffee plantations. Plant and Soil 67: 241-246.

Borst, G. 1986. Observations on a biological root rot control trial in the Fallbrook Area. California Avocado Soc. 70: 107-110.

Bos, L. 1981. Wild plants in the ecology of virus diseases. pp. 1-28. In: Maramorosch, K. and K. F. Harris. *Plant Diseases and Vectors: Ecology and Epidemiology.* Academic Press, New York.

Boserup, E. 1965. *The Conditions of Agricultural Growth: the Economics of Agrarian Change Under Population Pressure.* Allen and Unwin, London. 124 pp.

Boster, J. 1983. A comparison of the diversity of Jivaroan gardens with that of the tropical forest. Human Ecol. 11: 47-68.

______. 1984a. Classification, cultivation, and selection of Aguaruna cultivars of Manihot esculenta (Euphorbiaceae). Adv. in Econ. Bot. 1: 34-47.

______. 1984b. Inferring decision making from preferences and behavior: an analysis of Aguaruna Jivaro manioc selection. Human Ecol. 12: 343-358.

______. 1985. Selection for perceptual distinctiveness: evidence from Aguaruna cultivars of Manihot esculenta. Econ. Bot. 39: 310-325.

______. 1986. Exchange of varieties and information between Aguaruna manioc cultivators. Am. Anthropologist 88: 428-436.

Bouldin, D. R., S. W. Klausner and W. S. Reid. 1984. Use of nitrogen from manure. pp. 221-245. In: Hauck, R. D. ed., *Nitrogen in Crop Production.* Am. Soc. Agron., Crop Sci. Soc. Am., Soil Sci. Soc. Am., Madison, WI.

Boutelou, E. 1949. *Cultivo de la Vid en Jerez y Sanlucar.* Ser. A, Num. 2, Manuales Technicos. Min. Capacitación Agricultura y Propaganda, Madrid. 157 pp.

Bourke, R. M. 1985. Food, coffee, and casuarina: an agroforestry system in Papua New Guinea highlands. Agroforestry Systems 2: 273-279.

Bourne, M. C. 1977. Post harvest food losses -- the neglected dimension in increasing the world food supply. Cornell Intl. Agric. Mimeograph 53.

Ithaca, N. Y. 49 pp.

Boza, T. 1972. Ecological consequences of pesticides for the control of cotton insects in the Cañete Valley, Peru. pp. 423-438. In: Farvar, M. T. and J. P. Milton. eds., The Careless Technology: Ecology and International Development. Natural History Press, Garden City. 1030 pp.

Brass, L. J. 1941. Stone age agriculture in New Guinea. Geogr. Rev. 31: 555-569.

Bray , W. 1990. Agricultural renascence in the high Andes. Nature 345: 385.

Broadbent, S. M. 1987. The Chibcha raised-field system in the Sabana de Bogota, Colombia: further investigations. pp. 425-442. In: Denevan, W. M., K. Mathewson, and G. Knapp. Proc. 45th Intl. Congress of Americanists. Bogota, Colombia. BAR Intl. Series 359.

Brodie, B. B. 1982. Possible use of potato as a trap crop for controlling Globodera rostochiensis. J. Nematology 14: 432 .

______. 1984. Nematode parasites of potato. pp. 167-212. In: Nickle, W. ed., *Plant and Insect Nematodes.* Marcel Dekker, Basel.

Brodie, B. B. and W. F. Mai. 1989. Control of the golden nematode in the United States. Ann. Rev. Phytopathol. 27: 443-461.

Brodie, B. B. and W. S. Murphy. 1975. Population dynamics of plant nematodes as affected by combinations of fallow and cropping sequence. J. Nematology 7: 91-92.

Brokenshaw, D., and B. W. Riley. 1980. Mbeere knowledge of their vegetation and its relevance for development: a case-study from Kenya. pp. 111-127. In:

Brokenshaw, D., D. M. Warren and O. Werner eds., *Indigenous Knowledge Systems and Development.* Univ. Press of America, Washington, DC.

Brown, B. J. and G. G. Marten. 1986. The ecology of traditional pest management. pp. 241-272. In: Marten, G. G. ed., *Traditional Agriculture in Southeast Asia.* Westview Press, Boulder, CO.

Brown, L. N. 1933. Flooding to control root-knot nematodes. J. Agr. Res. 47: 883-888.

Brown, L. R. and P. Shaw. 1982. *Six Steps to a Sustainable Society.* Worldwatch Inst., Washington, D. C. 63 pp.

Brown, W. L. 1983. Genetic diversity and genetic vulnerability -- an appraisal. Econ. Bot. 37: 4-12.

Browning, J. A. 1975. Relevance of knowledge about natural ecosystems to development of pest management programs for agro-ecosystems. Proc. Am. Phytopathol. Soc. 1: 191-199.

______. 1981. The agro-ecosystem-natural ecosystem dichotomy and its impact on phytopathological concepts. pp. 159-172. In: Thresh, J. M. ed., *Pests, Pathogens, and Vegetation. The Role of Weeds and Wild Plants in the Ecology of Crop Pests and Diseases.* Pitman , Boston.

Browning, J. A. and K. J. Frey. 1969. Multiline cultivars as a means of disease control. Ann. Rev. Phytopathol. 7: 355-382.

Browning, J. A. and K. J. Frey. 1981. The multiline concept in theory and practice. pp. 37-46. In: J. F. Jenkyn and R. T. Plumb eds., *Strategies for the Control of Cereal Disease.* Blackwell Sci. Publ., Oxford.

Brush, S. B. 1977. Farming the edge of the Andes. Natural History 86(5): 32-40.

______. 1980. The environment and native Andean agriculture. Am. Indigena

25: 161-172.

______. 1981. Estrategias agrícolas tradicionales en las montañosas de America Latina. pp. 65-76. In: Novoa B., A. R. and J. L. Posner eds., *Seminario Internacional Sobre Produccion Agropecuaria y Forestal en Zonas de Ladera de America Tropical.* Informe Technico No. 11. CATIE, Turrialba, Costa Rica.

______. 1986. Genetic diversity and conservation in traditional farming systems. J. Ethnobiology 6: 151-167.

Brush, S. B., H. J. Carney, and Z. Huaman. 1981. Dynamics of Andean potato agriculture. Econ. Bot. 35: 70-88.

Buddenhagen, I. W. 1977. Resistance and vulnerability of tropical crops in relation to their evolution and breeding. pp. 309-326. In: Day, P. R. ed., *The Genetic Basis for Epidemics in Agriculture.* Ann. N. Y. Acad. of Sciences. Vol. 287.

______. 1981. Conceptual and practical considerations when breeding for tolerance or resistance. pp. 221-234. In: Staples, R. C. and G. H. Toenniessen eds., *Plant Disease Control . Resistance and Susceptibility.* Wiley, New York.

______. 1983. Breeding strategies for stress and disease resistance in developing countries. Ann. Rev. Phytopathol. 21: 385-409.

Buddenhagen, I. W. and T. A. Elasser. 1962. An insect spread bacterial wilt epiphytotic of Bluggoe banana. Nature 194: 164-165.

Buddenhagen, I. W. and L. Sequeira. 1958. Disinfectants and tool disinfestation for prevention of spread of bacterial wilt of bananas. Plant Dis. Reptr. 42: 1399-1404.

Bull, D. 1982. *A Growing Problem: Pesticides and the Third World Poor.* Oxfam, Oxford. 198 pp.

Bunch, Roland. 1982. *Two Ears of Corn.* World Neighbors, Oklahoma City, OK. 250 pp.

Burdon, J. J. 1978. Mechanisms of disease control in heterogeneous plant populations -- an ecologist's view. pp. 193-200. In: Scott, P. R. and A. Bainbridge eds., *Plant Disease Epidemiology.* Blackwell , Oxford.

Burdon, J. J. and G. A. Chilvers. 1982. Host density as a factor in plant disease ecology. Ann. Rev. Phytopathol. 20: 143-166.

Burke, D. W. 1964. Time of planting in relation to disease incidence and yields of beans in central Washington. Plant Dis. Reptr. 48: 789-793.

Butler, E. J. 1918. *Fungi and Diseases in Plants.* Thacker, Spink, Calcutta. 547 pp.

Butterfield, E. J., J. E. DeVay, and R. H. Garber. 1978. The influence of several crop sequences on the incidence of Verticillium wilt of cotton and on the population of Verticillium dahliae in field soil. Phytopathology 68: 1217-1220.

Calpouzos, L. 1966. Action of oil in the control of plant disease. Ann. Rev. Phytopathol. 4: 369-390.

Camino, A., J. Recharte and P. Bidegaray. 1981. Flexibilidad calendarica en la agricultura tradicional de las vertientes orientales de los Andes. pp. 169-194. In: *La Technologia en el Mundo Andino --I.* Mexico City, Mexico.

Campbell, L. 1949. Gray mold of beans in Western Washington. Plant Dis. Reptr. 33: 91-93.

Cancian, F. 1972. *Change and Uncertainty in a Peasant Community. The Mayan Corn Farmers of Zinacantan.* Stanford Univ. Press, Stanford, CA. 208 pp.

Cardenas-Alonso, M. R. 1989. Web blight of beans (Phaseolus vulgaris L.) incited by Thanatephorus cucumeris (Frank) Donk. in Colombia. Ph.D. Thesis, Cornell Univ., Ithaca, NY

Carment, A. G. 1922. Report on the mycological work done by Dr. A. G. Carment. pp. 8. In: Ann. Report 1921. Dept. of Agric., Legislative Council, Fiji.

Carneiro, R. L. 1988. Indians of the Amazonian forest. pp. 25-36. In: Denslow, J. S. and C. Padoch eds., *People of the Tropical Rain Forest.* Univ. of Calif. Press, Berkeley, CA.

Carr, A. ed. 1978. *The Encyclopedia of Organic Gardening.* Rodale Press, Emmaus, PA.

Carrier, L. 1923. *The Beginnings of Agriculture in America* . McGraw-Hill, New York. 323 pp.

Carter, W. E. 1969. *New Lands and Old Traditions. Kekchi Cultivators in the Guatemalan Lowlands.* Univ. of Florida Press, Gainesville, FL. 153 pp.

Carter, W. F. and M. Mamani. 1982. *Irpa Chico: individuo y comunidad en la cultura Aymara.* Ed. Juventud, La Paz.

Casas Gaspar, E. 1950. *Ritos Agrarios. Folklore Campesino Español.* Ed. Escelicer, Madrid. 310 pp.

Castillo, M. 1985. Some studies on the use of organic soil amendments for nematode control. Philippine Agriculturalist 68: 1-18.

Castillo, M. B., M. B. Arceo and J. A. Litsinger. 1978. Effect of geomorphic field position, flooding, and cropping pattern on plant parasitic nematodes of crops following rainfed wetland rice in Iloilo, Philippines. Intl. Rice Res. Newsletter 3 (5): 27.

Cato, Marco Porcius. 1934. *On Agriculture.* W. D. Hooper and H. B. Ash (trans.) Harvard Univ. Press, Cambridge. pp. 1-157.

Cavallini, R. 1972. Recommendaciones para aumentar la produción de frijol tapado. Agroindustria (Costa Rica) 1(6): 18.

Caveness, F. E. 1972a. Cambios en poblaciones de nematodos fitoparasitos en terreno recientemente limpiados. Nematropica 2(1): 15-16. (Abstr.).

______. 1972b. Changes in plant parasitic nematode populations on newly cleared land. Nematropica 2(1): 1-2.

______. 1976. Limited plant-parasitic nematode control with slash-and-burn farming. Nigerian Soc. Plant Protec. 6th Ann. General Conf. Univ. of Nigeria. Nsukka, Nigeria.

CGIAR. 1974. *International Research in Agriculture.* CGIAR, New York. 70 pp.

______. 1988. CGIAR 1987/88 Annual Report. CGIAR, Washington, D.C. 82 pp.

Chacón, J. C. & S. R. Gliessman. 1982. Use of the "non-weed" concept in traditional tropical agroecosystems of south-eastern Mexico. Agro-Ecosystems 8: 1-11.

Chambers, R., A. Pacey and L. A. Thrupp. 1990. Farmer First. Farmer Innovation and Agricultural Research. Intermediate Technology Publ., London. 219 pp.

Chan, P. 1985. *Better Vegetable Gardens the Chinese Way.* Garden Way Publ., Pownal, Vermont. 103 pp.

Chandler, R. F. 1981. Land and water resources and management. pp. 9-18. In: Plucknett, D.L. and H. L. Beemer, Jr, ed., 1981. *Vegetable Farming Systems in China.* Westview Press, Boulder, CO.

Chang, T. T. 1976. The origin, evolution, cultivation, dissemination and diversification of Asian and African rices. Euphytica 25: 425-441.

Chapin, G. and R. Wasserstrom. 1981. Agricultural production and malaria resurgence in Central America and India. Nature 293: 181-185.

Chapin, M. 1988. The seduction of models. Chinampa agriculture in Mexico. Grassroots Develop. 12 (1): 8-17.

Chen, A. 1987. Unraveling another Mayan mystery. Discover, June.

Chiu, W. F. and Y. H. Chang. 1982. Advances of science of plant protection in the People's Republic of China. Ann. Rev. Phytopathol. 20: 71-92.

Christanty, L. 1986. Shifting cultivation and tropical soils: Patterns, problems, and possible improvements. pp. 226-240. In: Marten, G. G. , ed.. *Traditional Agriculture in Southeast Asia.* Westview Press, Boulder, Co.

Christanty, L., O. S. Abdoellah, G. G. Marten, and J. Iskandar. 1986. Traditional Agroforestry in West Java: The Pekarangan (homegarden) and Kebun-Talun (annual-perennial rotation) cropping systems. pp. 132-158. In: Marten, G.G. ed., *Traditional Agriculture in Southeast Asia.* Westview Press, Boulder, CO.

Christensen, C. M. 1979. The effects of plant parasitic and other fungi on man. pp. 393-409. In: Horsfall, J. G. and E. B. Cowling, ed., *Plant Disease -- An Advanced Treatise. Vol. IV. How Pathogens Induce Disease.* Academic Press, New York.

Christensen, C. M. and H. H. Kaufmann. 1969. *Grain Storage. The Role of Fungi in Quality Loss.* Univ. Minnesota Press, Minneapolis. 153 pp.

Christiansen, J. A. 1977. The Utilization of Bitter Potatoes to Improve Food Production in the High Altitude of the Tropics. Ph.D. Thesis. Cornell Univ., Ithaca, NY 157 pp.

CIAT. 1975. Bean production systems program. Ser. FE - No. 5. CIAT, Cali, Colombia. 38 pp.

Cieza de León, Pedro de. 1959. *The Incas (Chronica del Peru).* Harriet de Onis (trans.) Victor W. von Hagen ed., Univ. of Oklahoma Press, Norman, OK. 397 pp.

______. 1984. *La Crónica del Peru.* Ballestros, M. ed., Historia 16, Madrid. 414 pp.

______. 1985. *El Señorio de los Incas.* Ballestros, M. ed., Historia 16, Madrid. 211 pp.

______. 1986. *Descubrimiento y Conquista del Peru.* De Santa Marta, C. S. ed., Historia 16, Madrid. 339 pp.

CIMMYT. 1974. *The Puebla Project: Seven Years of Experience: 1967-1973.* CIMMYT, El Batán, Mexico. 118 pp.

Clark, C. A. and J. W. Moyer. 1988. *Compendium of Sweet Potato Diseases.* Am. Phytopathol. Soc., St. Paul, MN. 74 pp.

Clavigero, F. J. 1974. *Historia Antigua de Mexico.* M. Cuevas. ed., Editorial Porrua, Mexico. 621 pp.

Clawson, D. L. 1985. Harvest security and intraspecific diversity in traditional tropical agriculture. Econ. Bot. 39: 56-67.

Clement, C. R. 1986. The pejibaye palm (Bactris gasipaes H. B.K.) as an agroforestry component. Agroforestry Systems 4: 205-219.

______. 1989. A center of crop genetic diversity in Western Amazonia. BioScience 39: 624-631.

Coe, M. D. 1964. The chinampas of Mexico. Sci. Am. 211: 90-98.

Coffey, M. D. 1984. An integrated approach to the control of avocado root rot. California Avocado Soc. 68: 61-68.

Coley-Smith, J. R. 1987. Alternative methods of controlling white rot diseases of Allium. pp. 161-177. In: Chet, I. *Innovative Approaches to Plant Disease Control.* Wiley, New York.

Columela, L. J. M. 1988. *De Los Trabajos de Campo.* R. A. Holgado ed., Min. Agric., Pesca y Alimentación. Siglo XXI de España Editores, Madrid. 339 pp.

Conklin, H. C. 1954. An ethnoecological approach to shifting agriculture. Trans. N.Y Acad. Sci. (Ser. 2) 17: 133-142.

______. 1957. *Hanunóo Agriculture: A Report on an Integral System of Shifting Cultivation in the Philippines.* FAO For. Develop. Paper No. 12. FAO, Rome. 209 pp.

______. 1961. The study of shifting cultivation. Curr. Anthropol. 2(1): 27-61.

______. 1980. *Ethnographic Atlas of Ifugao.* Yale Univ. Press, New Haven. 116 pp.

Conway, G. R. 1985. Agroecosystem analysis. Agric. Administration 20: 31-55.

______. 1986. *Agroecosystem Analysis for Research and Development.* Winrock Intl. 111 pp.

Cook, O. F. 1916. Staircase farms of the ancients. Nat. Geograph. 29: 474-534.

Cook, R. J. 1981. Fusarium diseases in the People's Republic of China. pp. 53-55. In: Nelson, P. E., T. A. Toussoun and R. J. Cook eds., *Fusarium Diseases, Biology and Taxonomy.* Pennsylvania State Univ. Press, University Park, PA.

______. 1986. Plant health and the sustainability of agriculture, with special reference to disease control by beneficial microorganisms. Biol. Agric. Hort. 3: 211-232.

Cook, R. J. and K. F. Baker. 1983. *The Nature and Practice of Biological Control of Plant Pathogens.* Am. Phytopathol. Soc., St. Paul, MN. 539 pp.

Cook, R. J., Boosalis, M. G. and B. Doupnik. 1978. Influence of crop residues on plant diseases. pp. 147-163. In: *Crop Residue Management Systems.* Am. Soc. Agron. Spec. Publ. 31. Madison, WI.

Coursey, D. G. 1967. *Yams.* Longman, London. 230 pp.

______. 1983. Post harvest losses in perishable foods of the developing world. pp. 485-514. In: Lieberman, M. ed., *Post-harvest Physiology and Crop Preservation.* Plenum, New York.

Coursey, D. G., and R. H. Booth. 1972. The post-harvest phytopathology of perishable tropical produce. Rev. Plant Pathology 51: 751-765.

Coursey, D. G., and F. J. Proctor. 1975. Towards the quantification of post-harvest loss in horticultural produce. Acta Horticulturae 49: 55-63.

Cowling, E. B. 1978. Agricultural and forest practices that favor epidemics. pp.

361-381. In: Horsfall, J. G. and E. B. Cowling, ed., *Plant Disease -- An Advanced Treatise. Vol. II. How Disease Develops in Populations.* Academic Press, New York.

Cralley, E. M. 1957. The effect of seeding methods on the severity of white tip of rice. Phytopathology 47: 7 (abstr.).

Cramer, H. H. 1967. *Plant Protection and World Crop Production.* Bayer, Leverkusen, Germany. 524 pp.

Crill, P. 1981. Twenty years of plant pathology at the IRRI. Plant Dis. 65: 569-574.

Crispin, A. and C. C. Gallegos. 1963. Web blight – a severe disease of beans and soybeans in Mexico. Plant Dis. Reptr. 47: 1010-1011.

Crosson, P. 1981. *Conservation Tillage and Conventional Tillage: a Comparative Assessment.* Soil Conservation Soc. Am., Ankeny, Iowa. pp. 35.

Curl, E. A. 1963. Control of plant disease by crop rotation. Bot. Rev. 29: 413-479.

Curwen, E. C. and G. Hatt. 1953. *Plough and Pasture. The Early History of Farming.* Collier, New York. 251 pp.

Dakwa, J. T. 1979. The effect of shade and NPK fertilizers on the incidence of cocoa black pod disease in Ghana. Ghana J. Agric. Sci. 9: 179-184.

D'Altroy, T. N. and T. K. Earle. 1985. Staple finance, wealth finance, and storage in the Inka political economy. Curr. Anthropology 26(2): 187-197.

D'Altroy, T. N. and C. Harstorf. 1984. The distribution and contents of Inca State storehouses in the Xauxa Region of Peru. Am. Antiquity 49 (2): 334-349.

Dalrymple, D. G. 1971. *Survey of Multiple Cropping in Less Developed Nations.* Foreign Econ. Dev. Report No. 12. Foreign Econ. Res. Serv., USDA/USAID, Washington, D.C. 108 pp.

______. 1978. *Development and Spread of High-Yielding Varieties of Wheat and Rice in the Less Developed Nations.* Foreign Agric. Econ. Report. No. 95. Econ. Res. Serv., USDA, Washington, DC.

Daly, H. E. ed. 1980. *Economics, ecology, ethics: essays toward a steady-state economy.* Freeman, San Francisco. 372 pp.

Darch, J. P. ed. 1983. *Drained Field Agriculture in Central and South America.* BAR Intl. Ser. 189. Oxford. 263 pp.

Davies, J. C. 1976. The incidence of rosette disease in groundnut in relation to plant density and it effects on yield. Ann. Appl. Biol. 82: 489-501.

Davis, G. 1988. The Indonesian transmigrants. pp. 143-153 In: Denslow, J. S. and C. Padoch eds., *People of the Tropical Rain Forest.* Univ. of Calif. Press, Berkeley, CA.

Day, P. R. 1977. *The Genetic Basis for Epidemics in Agriculture.* Ann. N. Y. Acad. Sci. Vol. 287. 400 pp.

Dazhong, W. and D. Pimentel. 1986. Seventeenth Century organic agriculture in China. I. Cropping systems in the Jiaxing region. Human Ecol. 14: 1-14.

De Acosta, J. 1987. *Historia Natural y Moral de las Indias.* J. Aleina Franch ed., Historia 16, Madrid. 515 pp.

De Asso, I. 1947. *Historia de la Economía Politica de Aragon.* Consejo Superior de Investigaciones Cientificas. Zaragoza, Spain.

De Bokx, J. A. and J. P. H. van der Want eds., 1987. *Viruses of Seed Potatoes and Seed-Potato Production.* Pudoc, Wageningen. 259 pp.

De Datta, S. K. 1981. *Principles and Practices of Rice Production.* Wiley, New York. 618 pp.

De Datta, S. K., and R. Q. Lacsina. 1972. Weed control in flooded rice in tropical Asia. pp. 472-478. In: Proc. 11th Br. Weed Control Conf., Brighton, England.

De el Seixo, V. 1793. *Lecciones Prácticas de Agricultura y Economia.* Vol. 3. Pantaleon Aznar, Madrid. 299 pp.

De la Vega, Garcilaso, El Inca. 1966. *Royal Commentaries of the Incas and General History of Peru. Part 1.* H. V. Livermore (trans.) Univ. Texas Press, Austin. 627 pp.

De Landa, Diego. 1985. *Relación de las Cosas de Yucatán.* M. Rivera ed., Historia 16, Madrid. 187 pp.

Del Busto, J. A. 1978. *Peru Incaico.* Liberia Studium, Lima. 385 pp.

De Murúa, Martin. 1987. *Historia General de Peru.* Historia 16, Madrid. 583 pp.

De Sahagún, Friar Bernadino. 1969. *Historia General de las Cosas de Nueva España.* 4 Vol. Editorial Porrua, Mexico.

De Schlippe, P. 1956. *Shifting Cultivation in Africa.* Routledge and Kegan, London. 304 pp.

De Shield, M. A. 1962. Shifting Cultivation in the Tropics with Particular Emphasis on Liberia. M.S. Thesis, Cornell Univ., Ithaca, NY. 93 pp.

De Torquemada, Juan 1969. *Monarquia Indiana.* M. L. Portilla. ed., Editorial Porrua, Mexico.

Denevan, W. M. 1970. Aboriginal drained-field cultivation in the Americas. Science 169: 647-654.

______. 1985. Peru's agricultural legacy. Focus. Univ. Wisconsin, Madison, WI. April. pp. 16-29.

______. 1987. Terrace abandonment in the Colca Valley, Peru. pp. 1-43. In: Denevan, W. M., K. Mathewson and G. Knapp eds., Pre-hispanic Agricultural Fields in the Andean Region. BAR Intl. Ser. 359 (i). Oxford.

Denevan, W. M. and C. Padoch. eds. 1987. *Swidden-fallow Agroforestry in the Peruvian Amazon.* New York Bot. Garden, Bronx, NY. 107 pp.

Denevan, W. M. and J. M. Treacy. 1987. Young managed fallow at Brillo Nuevo. pp. 8-46. In: Denevan, W. M. and C. Padoch, ed., *Swidden-fallow Agroforestry in the Peruvian Amazon.* New York Bot. Garden, New York.

Denevan, W. M. and B. L. Turner II. 1974. Forms, function, and associations of raised fields in the old world tropics. J. Trop. Geography 39: 24-33.

Denevan, W. M., K. Mathewson, and G. Knapp. 1987. *Pre-hispanic Agricultural Fields in the Andean Region.* Proc. 45th Intl. Congress of Americanists, Bogota, Colombia. BAR Intl. Series 359, Oxford. 504 pp.

Dennis Jr., J. V. 1987. Farmer Management of Rice Variety Diversity in Northern Thailand. Ph.D. Thesis. Cornell Univ., Ithaca, NY. 367 pp.

Denslow, J. S. 1988. The tropical rain-forest setting. pp. 25-36. In: Denslow, J. S. and C. Padoch eds., *People of the Tropical Rain Forest.* Univ. of Calif. Press, Berkeley, CA.

Dickson, J. G. 1947. *Diseases of Field Crops.* McGraw-Hill, New York. 429 pp.

Dies, E. J. 1949. *Titans of the Soil. Great Builders of Agriculture.* Univ. of North Carolina Press, Chapel Hill, NC. 213 pp.

D'Olwer, L. N. 1963. *Cronistas de las Culturas Precolombinas.* Fondo Cultura Económica, Mexico. 756 pp.

Dominguez Garcia-Tejero, F. 1989. *Plagas y Enfermedades de las Plantas Cultivadas.* Ediciones Mundi-Prensa, Madrid. 821 pp.

Donkin, R. A. 1979. *Agricultural Terracing in the Aboriginal New World.* Univ. Arizona Press, Tucson. 196 pp.

Doupnik Jr., B., M. G. Boosalis, G. Wicks, and D. Simka. 1975. Ecofallow reduces stalk rot in grain sorghum. Phytopathology 65: 1021-1022.

Dove, M. R. 1983. Theories of swidden agriculture, and the political economy of ignorance. Agroforestry Systems 1: 85-99.

Dover, M. J. and L. M. Talbot. 1987. *To Feed the Earth: Agro-ecology for Sustainable Development.* World Resources Inst., Washington, DC. 88 pp.

Duffus, J. E. 1971. Role of weeds in the incidence of virus diseases. Ann. Rev. Phytopathol. 9: 319-340.

Dykstra, T. P. 1961. Production of disease-free seed. Bot. Rev. 27: 445-500.

Echandi, E. 1965. Basidiospore infection by Pellicularia filamentosa (= Corticum microsclerotia), the incitant of web blight of the common bean. Phytopathology 55: 698-699.

Eckholm, E. P. 1976. *Losing Ground. Environmental Stress and World Food Prospects.* Norton, New York. 223 pp.

Eden, M. J. 1988. Crop diversity in tropical swidden cultivation: comparative data from Colombia and Papua New Guinea. Agric., Ecosystems and Environment 20(2): 127-136.

Eden, M. J. and A. Andrade. 1987. Ecological aspects of swidden cultivation among the Andoke and Witoto Indians of the Colombian Amazon. Human Ecol. 15: 339-360.

Edwards, C. A., R. Lal, P. Madden, R. H. Miller and G. House. eds. 1990. *Sustainable Agricultural Systems.* Soil & Water Conserv. Soc., Ankeny, Iowa. 696 pp.

Ekundayo, J. A. and S. H. Z. Naqvi. 1972. Preharvest microbial rotting of yams (Dioscorea spp.) in Nigeria. Trans. Br. Mycological Soc. 58: 15-18.

Emerson, R. A. 1953. A preliminary survey of the milpa system of maize culture as practiced by the Maya Indians of the Northern part of the Yucatan peninsula. Ann. Missouri Bot. Garden 40: 51-62.

Ene, L. S. O. 1977. Control of cassava bacterial blight (CBB). Trop. Root and Tuber Crops Newsletter 10: 30-31.

Equaras Ibañez, J. ed. 1988. *Ibn Luyun: Tratado de Agricultura.* Patronato de la Alahambra y Generalife, Granada, Spain.

Erickson, C. L. 1985. Applications of prehistoric Andean technology experiments in raised field agriculture, Huatta, Lake Titicaca: 1981-1982. pp. 209-232. In: Farrington, I.S. ed., *Prehistoric Intensive Agriculture in the Tropics.* BAR Intl. Ser.232, Oxford.

Erickson, C. L. and K. L. Candler. 1989. Raised fields and sustainable agriculture in the Lake Titicaca basin of Peru. pp. 230-249. In: Browder, J. O. ed., *Fragile Lands of Latin America: Strategies for Sustainable Development.* Westview, Boulder, CO.

Estrada R., N., E. Perez M., and L. Heidrick. 1959. DICOL Monserrate, una

Nueva Variedad de Papa. Dept. de Investigación Agropecuaria, Boletin de Divulgacion No. 6. Bogotá, Colombia.

Estrella, E. 1986. *El Pan de America. Ethnohistoria de los Alimentos Aborígenes en el Ecuador.* Consejo Superior Investigaciones Cientificas, Centros Estudios Historicos, Madrid. 390 pp.

Evans, A. C. 1960. Studies of intercropping. Part I. Maize or sorghum with groundnuts. E. Afr. Agric. For. J. 26: 1-10.

Ewel, J., C. Berish, B. Brown, N. Price, and J. Raich. 1981. Slash and burn impacts on a Costa Rican wet forest site. Ecology 62: 816-829.

Ewell, P. T. and T. T. Poleman. 1980. *Uxpanapa: Agricultural Development in the Mexican Tropics.* Pergamon, New York. 207 pp.

Ewell, P. T. and D. Merrill-Sands. 1987. Milpa in Yucatán: a long-fallow maize system and its alternatives in the Maya peasant economy. pp. 95-129. In: Turner, B. L. and S. B. Brush, ed., *Comparative Farming Systems.* Guilford Press, New York.

Ewell, P. T., H. Fano, K. V. Raman, J. Alcázar, M. Palacios, and J. Carhuamaca. 1990. Farmer Management of Potato Insect Pests in Peru. Intl. Potato Center, Lima, Peru. 87 pp.

Fagan, H. J. 1987. Influence of microclimate on Drechslera leaf spot of young coconuts: effect of desiccation on spore survival and of moisture and shade on infection and disease development. Ann. Appl. Biol. 111: 521-533.

Fallers, M. C. 1960. *The Eastern Lacustrine Bantu (Ganda and Soga).* Intl. Afr. Inst. 86 pp. Ethnographic Survey of Africa: East Central Africa, pt. 11.

FAO. 1970. *World Census of Agriculture.* FAO, Rome. 289 pp.

______. 1980. *China: Multiple Cropping and Related Crop Production Technology.* Plant Production and Protec. Paper 22. FAO, Rome. 57 pp.

______. 1986. *FAO Production Yearbook.* Food and Agric. Organ., Rome. 309 pp.

Farrell, J. A. K. 1976. Effects of intersowing with beans on the spread of groundnut rosette virus by Aphis craccivora (Hemiptera, Aphididae) in Malawi. Bull. Entomol. Res. 66: 331-333.

Fernandes, E. C. M., A. Oktingati, and J. Maghembe. 1984. The Chagga homegardens: a multistoried agroforesty cropping systems on Mt. Kilimanjaro (Northern Tanzania). Agroforestry Systems 2: 73-86.

Fernandez de Oviedo, G. 1986. *Sumario de la Natural Historia de las Indias.* M. Ballesteros ed., Historia 16, Madrid. 181 pp.

Finegan, E. J. 1981. The Use of Agri-silviculture as a Resource Conservation and Rural Community Development Method in the Tropical Wet Forest of Colombia. Ph.D. Thesis, Cornell Univ., Ithaca, NY.

Fitt, B. D. L. and H. A. McCartney. 1986. Spore dispersal in splash droplets. pp. 87-104. In: Ayres, P. G. and L. Boddy, ed., *Water, Fungi and Plants.* Cambridge Univ. Press, Cambridge.

Flores Ochoa, J. A. 1987. Evidence for the cultivation of Qochas in the Peruvian altiplano. pp. 399-402. In: Denevan, W. M., K. Mathewson, and G. Knapp. Proc. 45th Int. Congress of Americanists. Bogota, Colombia. British Archelogical Reports, Int. Series 359. BAR, Oxford.

Flouri, F., I. Chatjipavlidis, C. Balis, D. Servis and C. Tjerakis. 1990. Effect of olive oil mills liquid wastes on soil fertility. Reunion Int. Sobre Tratamiento de Alpechines, Cordoba, Spain. May 31-June 1, 1990. Consejo Oleicola Internacional. Mimeo. 11 pp.

Forcella, F. 1988. Importance of pesticide alternatives to sustainable agriculture. pp. 41-43. Nat. Forum. Phi Kappa Phi J. Summer 1988.

Fowler, M. L. 1969. Middle Mississippian agricultural fields. Am. Antiquity 34: 365-375.

Fox, R. A. 1970. A comparison of methods of dispersal, survival, and parasitism in some fungi causing root diseases of tropical plantation crops. pp. 179-187. In: Tousson, T. A., R. V. Bega, and R. E. Nelson eds., *Root Diseases and Soil-borne Pathogens.* Univ. Calif. Press, Berkeley.

Francis, C. A. ed. 1986. *Multiple Cropping Systems*. Macmillan, New York. 383 pp.

______. 1988. Biological efficiencies in multiple-cropping systems. Adv. Agron. 42: 1-42.

Francis, C. A., C. Butler Flora and L. D. King. 1990. *Sustainable Agriculture in Temperate Zones.* Wiley, New York. 487 pp.

Frank, J. A. and H. J. Murphy. 1977. The effect of crop rotations on Rhizoctonia disease of potatoes. Am. Potato J. 54: 315-322.

Frankel, O. H. 1970. Genetic conservation in perspective. pp. 469-489. In: Frankel, O. H. and E. Bennett, ed., *Genetic Resources in Plants.* Blackwell Sci., Oxford.

Fresco, L. O. 1986. *Cassava in Shifting Cultivation.* Royal Trop. Inst., Amsterdam. 240 pp.

Fritz, V. A. and S. Honma. 1987. The effect of raised beds, population densities, and planting date on the incidence of bacterial soft rot in Chinese cabbage. J. Am. Soc. Hort. Sci. 112: 41-44.

Fry, W. E. 1982. *Principles of Plant Disease Management.* Academic Press, New York. 378 pp.

Gade, D. W. 1975. *Plants, man and the land in the Vilcanota valley of Peru.* W. Junk, The Hague. 240 pp.

Galindo, J. J. 1982. Epidemiology and control of web blight of beans in Costa Rica. Ph.D. Thesis. Cornell Univ., Ithaca, NY. 141 pp.

______. 1987. La moniliasis del cacao en centro America. pp. 7-16. In: Pinochet, J. ed., *Plagas y Enfermedades de Caracter Epidemico en Cultivos Frutales de la Region Centroamericana.* CATIE, Proyecto MIP, Panama. Informe Technico No. 110.

Galindo, J. J., G. S. Abawi, H. D. Thurston and G. Galvez. 1982. "Tapado", controlling web blight of beans on small farms in Central America. NY Food and Life Sciences 14: 21-25.

Galindo, J. J., G. S. Abawi, H. D. Thurston and G. Galvez. 1983a. Source of inoculum and development of bean web blight in Costa Rica. Plant Dis. 67: 1016-1021.

Galindo, J. J., G. S. Abawi, H. D. Thurston and G. Galvez. 1983b. Effect of mulching on web blight of beans in Costa Rica. Phytopathology 73: 610-615.

Garatcochea Z., I. 1985. Potencial agricola de los camellones en el altiplano Puneño. pp. 241-251 In: *Andenes y Camellones en el Peru Andino: Historia*

Presente y Futuro. CONCYTEC (Consejo Nacional de Ciencia y Technologia), Lima, Peru. 379 pp.

______. 1987. Agricultural experiments in raised fields in the Lake Titicaca basin, Peru: preliminary considerations. pp. 385-398. In: Denevan, W. M., K. Mathewson, and G. Knapp, ed., Proc. 45th Intl. Congress of Americanists. Bogota, Colombia. BAR Intl. Series 359 (ii). Oxford.

Garcia-Badell y Abadia, G. 1963. *Introdución a la Historia de la Agricultura Española.* Consejo Superior de Investigaciones Cientificas, Madrid.

Garcia-Espinosa, R. 1980a. Chenopodium ambrosoides L., planta con uso potencial en el combate de nematodos fitoparasitos. Agricultura Trop. (CSAT) 2: 92-97.

______. 1980b. Incidencia de algunos fitopatógenos del suelo en maiz, con enfasis en Pythium sp. bajo dos tipos distinto de agroecosistemas en el tropico humedo. Agricultura Trop. (CSAT) 2: 98-104.

______. 1987. Importancia de la fitopatologia tropical. pp. 1-17. In: *Taller de Fitopatologia Tropical.* 2nd Edition. Colegio de Postgraduados, Chapingo, Mexico.

Garcia-Rodriguez, A. 1990. Eliminación y aprovechamiento agricola del alpechin. Reunion Int. Sobre Tratamiento de Alpechines, Cordoba, Spain. May 31-June 1, 1990. Consejo Oleicola Internacional. Mimeo. 35 pp.

Garrett, S. D. 1960. *Biology of Root-infecting Fungi.* Cambridge Univ. Press. 293 pp.

Gäumann, E. 1950. *Principles of Plant Infection.* Hafner, New York. 543 pp.

Geertz, C. 1963. *Agricultural Involution.* Univ. of California Press, Berkeley. 176 pp.

Gilbert, J. C. 1956. Soil mulches of local material. Hawaii Farm Sci. 4 (4): 4-5.

Gill, M. A. and J. M. Bonman. 1988. Effects of water deficit on rice blast I. Influence of water deficit on components of resistance. J. Plant Protec. Tropics 5 (2): 61-66.

Gindrat, D. 1979. Biological soil disinfestation. pp. 253-287. In: Mulder, D. ed., *Soil Disinfestation.* Elseveir, Amsterdam.

Glass, E. H. and H. D. Thurston. 1978. Traditional and modern crop protection in perspective. BioScience 28: 381-383.

Gleissman, S. R. 1986. Plant interactions in multiple cropping systems. pp. 82-95. In: Francis, C. A. ed., 1986. *Multiple Cropping Systems.* Macmillan, New York.

______. 1988. The home garden agroecosystem: a model for developing sustainable tropical agricultural systems. pp. 445-450. In: Allen, P. and D. van Dusen, ed., *Global Perspectives on Agroecology and Sustainable Agricultural Systems.* Vol. 2. Agroecology Program, Univ. of California, Santa Cruz, CA.

Glover, N. and J. Beer. 1986. Nutrient cycling in two traditional Central American agroforestry systems. Agroforestry Systems 4: 77-87.

Glynne, M.D. 1965. Crop sequence in relation to soil-borne pathogens. pp. 423-435. In: Baker, K. F. and W. C. Synder. *Ecology of soil-borne pathogens.* Univ. of Calif. Press, Berkeley, CA.

Godfrey, G. H. and H. M. Hoshino. 1934. The trap crop as a means of reducing root-knot nematode infestation. Phytopathology 24: 635-658.

Gomez, A. A., and K. A. Gomez. 1983. *Multiple Cropping in the Humid Tropics of Asia.* IDRC 176e. IDRC, Ottawa, Canada. 248 pp.

Gomez, P. L. and D. E. van de Zaag. 1986. The role of the national potato programme in Colombia. Potato Res. 29: 245-250.

Gómez-Pompa, A. 1978. An old answer to the future. Mazingra 5: 51-55.

Gonzalez J., A. 1985. Home gardens in Central Mexico. pp. 521-537. In: Farrington, I.S. ed., *Prehistoric Intensive Agriculture in the Tropics.* BAR Intl. Ser. 232, Oxford.

Good, J. M. 1968. Relation of plant parasitic nematodes to soil management practices. pp. 113-138. In: Smart Jr., G. C. and V. G. Perry eds., *Tropical Nematology.* Univ. Florida Press, Gainesville, FL.

Goodall, G. E. , H. D. Ohr, and G. A. Zentmeyer. 1987. Mounds aid root rot replants. California Avocado Soc. 71: 147-151.

Goodell, G. 1984. Challenges to international pest management research and extension in the third world: do we really want IPM to work? Bul. Ent. Soc. Am. 30: 18-26.

Gourou, P. 1966. *The Tropical World.* Wiley, New York. 196 pp.

Grainge, M. and S. Ahmed. 1988. *Handbook of Plants with Pest-Control Properties.* Wiley, New York. 470 pp.

Granados A., N., R. Garcia E., E. Zavaleta M., R. Ferrera C., A. Castillo M., I. C. del Prado V. y P. Rodriguez G. 1990. Perdidas de grano por fitopatógenos del suelo en maiz monocultivo y rotado con leguminosas de coberatura en Tabasco, Mexico. Revista Mexicana de Fitopatologia. In Press.

Greaney, F. J. 1946. Influence of time, rate and depth of seeding on the incidence of root rot of wheat. Phytopathology 36: 252-263.

Greenland, D. J. 1975. Bringing the green revolution to the shifting cultivator. Science 190: 841-844.

Grigg, D. B. 1974. *The Agricultural Systems of the World: An Evolutionary Approach.* Cambridge Univ. Press, Cambridge. 358 pp.

Grist, D. H. 1975. *Rice*. Longman, London. 548 pp.

Grogan, R. G. and K. A. Kimble. 1967. The role of seed contamination in the transmission of Pseudomonas phaseolicola in Phaseolus vulgaris. Phytopathology 57: 28-31.

Guillet, D. 1987. Terracing and irrigation in the Peruvian highlands. Curr. Anthropology 28: 409-430

Gunn, D. L. and J. G. R. Stevens eds. 1976. *Pesticides and Human Welfare.* Oxford Univ. Press, Oxford. 278 pp.

Guthrie, J. W., L. L. Dea, C. L. Butcher, H. S. Fenwick, and A. M. Finley. 1975. The epidemiology and control of halo blight in Idaho. Idaho Agric. Exp. Sta. Bull. No. 550, Univ. of Idaho, Moscow, ID. 11 pp.

Guzman, V. L., et al. 1973. Celery production on organic soils of South Florida. Fl. Agric. Exp. Sta. Bull. 757. Univ. of Florida, Gainesville. 79 pp.

Hahn, S. K., D. S. O. Osiru, M. O. Akoroda and J. A. Otoo. 1987. Yam production and its future prospects. Outlook on Agriculture 16: 105-110.

Hall, D. W. 1969. Food storage in the developing countries. Trop. Sci. 11: 298-

318.
Hall, I. V., L. E. Aalders, L. P. Jackson, G. W. Wood, C. L. Lockhart. 1972. Lowbush blueberry production. Can. Dept. Agric. Publ. 1477. 42 pp.
Hammerton, J. L. 1985. Weed control in small farm systems. Proc. Caribbean Food Crops Soc., v. 20, St. Croix, VI. pp. 133-136.
Hames, R. B. 1983. Monoculture, polyculture and polyvariety in tropical forest swidden cultivation. Human Ecol. 2:13-34.
Han, Sun. 1987a. Water resources – the lifeline of Chinese agriculture. pp. 49-63. In: Wittwer, S., Y. Youtai, S. Han and W. Lianzheng, ed., *Feeding a Billion: Frontiers of Chinese Agriculture.* Michigan State Univ. Press, Lansing, MI.
______. 1987b. 9.6 million kilometers of territory. pp. 35-47. In: Wittwer, S., Y. Youtai, S. Han and W. Lianzheng, ed., *Feeding a Billion: Frontiers of Chinese Agriculture.* Michigan State Univ. Press, Lansing, MI.
Hardison, J. R. 1976. Fire and flame for plant disease control. Ann. Rev. Phytopathol. 14: 355-379.
Harlan, J. R. 1971. Agricultural origins, centers and noncenters. Science 174: 468-474.
______. 1975. Our vanishing genetic resources. Science 188: 618-621.
______. 1976. Diseases as a factor in plant evolution. Ann. Rev. Phytopathol. 14: 31-51
Harris, D. R. 1971. The ecology of swidden cultivation in the upper Orinoco rain forest, Venezuela. Geograph. Rev. 61: 475-495.
______. 1972. The origins of agriculture in the tropics. Am. Scientist 60: 180-193.
Harrison, P. D. and B. L. Turner II eds. 1978. *Pre-Hispanic Maya Agriculture.* Univ. New Mexico Press, Albuquerque, NM. 414 pp.
Hart, R. D. 1980. A natural ecosystem analog approach to the design of a successional crop system for tropical forest environments. Biotropica 12: 73-82. (Supplement on Tropical Succession).
Harwood, R. R. 1979. *Small Farm Development: Understanding and Improving Farming Systems in the Humid Tropics.* Westview, Boulder, CO. 160 pp.
Harwood, R. R. and D. L. Plucknett. 1981. Vegetable cropping systems. pp. 45-118. In: Plucknett, D. L. and H. L. Beemer, Jr, ed., *Vegetable Farming Systems in China.* Westview Press, Boulder, CO.
Haskell, P. T., T. Beacock and P. J. Wortley. 1981. World-wide socio-economic constraints to crop production. pp. 39-41. In: Kommedahl, T. ed., Proc. Symp. IX Intl. Congress of Plant Protection, Vol. 1., Washington, DC.
Hatch, J. K. ed. 1983. Our Knowledge. Traditional Farming Practices in Rural Bolivia. Vol. 1. Altiplano Region. Rural Develop. Services, New York. 337 pp.
Hauck, F. W. 1974. Shifting cultivation and soil conservation in Africa. FAO Soil Bull. 24: 1-4.
Have, H. and H. E. Kaufmann. 1972. Effect of nitrogen and spacing on bacterial leaf blight of rice. Indian Farming 22 (10): 8-13.
Hawkes, J. G. 1983. *The Diversity of Crop Plants.* Harvard Univ. Press, Cambridge, MA. 184 pp.
Hayes, R. T. 1932. Groundnut rosette disease in the Gambia. Trop. Agric.

(Trinidad) 9: 211-217.

Hayes, W. H. 1982. *Minimum tillage farming.* No-till Farmer, Inc., Brookfield, WI. 167 pp.

Herdt, R. W. and C. Capule. 1983. *Adoption, Spread, and Production Impact of Modern Varieties in Asia.* IRRI, Los Banos, Philippines. 54 pp.

Herklots, G. A. C. 1972. *Vegetables in South-east Asia.* Hafner, New York. 525 pp.

Hernandez X., E. 1985. Graneros de maiz en Mexico. pp. 205-230. In: Hernandez Z., E. Xolocotzia. *Tomo I & II.* Univ. Autonoma Chapingo, Chapingo, Mexico.

Hilton, R. N. 1952. Bird's eye spot disease of Hevea rubber caused by Helminthosporium hevea Petch. J. Rubber Res. Inst. Malaya 14: 40-82.

Hiraoka, M. 1989. Agricultural systems on the floodplains of the Peruvian Amazon. pp. 75-101. In: Browder, J. O. ed., *Fragile Lands of Latin America.* Westview, Boulder, CO.

Hirst, J. M. and O. J. Stedman. 1960. The epidemiology of Phytophthora infestans. I. Climate, ecoclimate and the phenology of disease outbreaks. Ann. Appl. Biol. 48: 471-488.

Hoitink, H. A. J. and P. C. Fahy. 1986. Basis for the control of soilborne plant pathogens with composts. Ann. Rev. Phytopathol. 24: 93-114.

Holliday, P. 1989. *A Dictionary of Plant Pathology.* Cambridge Univ. Press, Cambridge. 369 pp.

Hollings, M. 1965. Disease control through virus-free stock. Ann. Rev. Phytopathol. 3: 376-389.

Hollis, J. P. and T. Johnson. 1957. Microbiological reduction of nematode populations in water saturated soils. Phytopathology 47: 16 (Abstr.)

Hooper, D. J. and A. R. Stone. 1981. Role of wild plants and weeds in the ecology of plant parasitic nematodes. pp. 199-215. In: Thresh, J. M. ed., *Pests, Pathogens, and Vegetation. The Role of Weeds and Wild Plants in the Ecology of Crop Pests and Diseases.* Pitman, Boston.

Hopf, M. 1957. *Botanik un Vorgeschichte.* Jahrb. Rom. Germ. Museum. Mainz, Germany. pp. 1-27.

Hornby, D. 1983. Suppressive soils. Ann. Rev. Phytopathol. 21: 65-85.

Horsfall, J. G. 1945. *Fungicides and Their Action.* Chronica Botanica, Waltham, MA. 239 pp.

Hsu, Cho-yun. 1980. *Han Agriculture.* Univ. Washington Press, Seattle, WA. 377 pp.

Huang, H. T. and Pei Yang. 1987. The ancient cultured citrus ant. BioScience 37 (9): 655-671.

Huapaya, F., B. Salas, and L. Lescano. 1982. Ethnophytopathology of the Aymara communities of the Titicaca Lake Shore. Fitopatologia 17: 8.

Huber, D. M. and R. D. Watson. 1970. Effect of inorganic enrichment on soil-borne plant pathogens. Phytopathology 60: 22-26.

Hughes, E. 1973. Pink death in Iraq. The Reader's Digest. November.

Hurt, R. D. 1987. *Indian Agriculture in America. Prehistory to the Present.* Univ. Press of Kansas, Lawrence, KA. 290 pp.

ICRISAT. 1981. Proceedings of the International Workshop on Intercropping for

the Semi-arid Tropics. ICRISAT, Hyderbad, India. 401 pp.
IITA. 1975. Annual Report for 1974. IITA, Ibadan, Nigeria.
______. 1976. Annual Report for 1975. IITA, Ibadan, Nigeria.
______. 1977. Annual Report for 1976. IITA, Ibadan, Nigeria.
______. 1988. Annual Report for 1987. IITA, Ibadan, Nigeria.
Imle, E. P. 1978. Hevea rubber -- past and future. Econ. Bot. 32: 264-277.
Ioannou, N. , R. W. Schneider, and R. G. Grogan. 1977. Effect of flooding on the soil gas composition and the production of microsclerotia of Verticillium dahliae in the field. Phytopathology 67: 651-656.
IRRI. 1973. Annual Report for 1972. IRRI, Los Baños, Philippines.
______. 1974. Annual Report for 1973. IRRI, Los Baños, Philippines.
______. 1978. Annual Report for 1977. IRRI, Los Baños, Philippines.
______. 1979. Annual Report for 1978. IRRI, Los Baños, Philippines.
______. 1988. Annual Report for 1987. IRRI, Los Baños, Philippines.
______. 1990. Annual Report for 1989. IRRI, Los Baños, Philippines.
Jameson, J. D. ed. 1970. *Agriculture in Uganda.* Uganda Govt. Min. of Agric. For., Oxford Univ. Press, London. 395 pp.
Jeavons, J. 1982. *How To Grow More Vegetables.* Ten Speed Press, Berkeley, CA. 159 pp.
Jeffrey, D. 1974. Burma's leg rowers and floating farms. Natl. Geographic 145: 827-845.
Jeger, M. J., E. Griffiths and D. G. Jones. 1981. Effects of cereal cultivar mixtures on disease epidemics caused by splash dispersed pathogens. pp. 81-88. In:
Jenkyn, J. F. and R. T. Plumb. 1981. *Strategies for the Control of Cereal Disease.* Blackwell Sci. Publ., Oxford. 219 pp.
Jennings, P. 1976. The amplification of agricultural production. Sci. Am. 235(3): 180-194.
Jensen, N.F. 1952. Intra-varietal diversification in oat breeding. Agron. J. 44: 30-34.
Jimenez, S. E. 1978. Comentarios sobre la produción de frijol comun (Phaseolus vulgaris L.) en Costa Rica. Agron. Costarricense 2: 103-108.
Jiménez-Osornio, J. J. and S. del Amo R. 1988. An intensive Mexican traditional agroecosystem: the chinampa pp. 451-455 In: Allen, P. and D. van Dusen. eds., *Global Perspectives on Agroecology and Sustainable Agricultural Systems.* Vol. 2. Agroecology Prog., Univ. California, Santa Cruz, CA.
Jodha, N. S. 1979. Intercropping in traditional farming systems. pp. 282-291. In: ICRISAT. Proc. of the International Workshop on Intercropping for the Semi-arid Tropics. ICRISAT, Hyderbad, India.
Johns, T. and G. H. N. Towers. 1981. Isothiocyanates and thioreas in enzyme hydrolisis of Tropaeolum tuberosum. Phytochemistry 20: 2687-2689.
Johns, T., W. D. Kitts, F. Newsome and G. H. N. Towers. 1982. Anti-reproductive and other medicinal effects of Tropaeolum tuberosum. .J. Ethnopharmacology 5: 149-161.
Johns, T. and J. O. Kokwaro. 1991. Food plants of the Luo of Siaya district. Econ. Bot. 45: 103-113.
Johnston, S. A. and J. K. Springer. 1977. Pepper. Phytophthora blight cultural

control test . Plant Pathol. Leaflet 104. Rutgers State Univ., New Jersey.

Johnson, S. R. and R. D. Berger. 1972. Nematode and soil fungi control in celery seedbeds on muck soil. Plant Dis. Reptr. 56: 661-664.

Johnston, V. R. 1970. The ecology of fire. Audubon 72 (5): 76-119.

Jones, F. G. W. 1970. The control of the potato-cyst nematode. R. Soc. Arts 117: 179-199.

Jones, F. G. W. 1972. Management of nematode populations in Great Britain. Proc. Tall Timbers Conf. Feb. 24-25, 1972. Tall Timbers Res. Sta., Tallahassee, FL. pp. 81-107.

Jones, G. H. and A. E. G. Sif-El-Nasr. 1940. The influences of sowing depth and moisture on smut diseases, and the prospects of a new method of control. Ann. Appl. Biol. 27: 35-57.

Jones, R. A. C. 1981. The ecology of viruses infecting wild and cultivated potatoes in the Andean region of South America. pp. 89-107. In: Thresh, J.M. ed., *Pests, Pathogens, and Vegetation. The Role of Weeds and Wild Plants in the Ecology of Crop Pests and Diseases.* Pitman, Boston.

Jones, W. O. 1959. *Manioc in Africa.* Stanford Univ. Press, Stanford, CA. 315 pp.

Julien, C. J. 1985. Guano and resource control in sixteenth-century Arequipa. pp. 185-231. In: Masuda, S., I. Shimada and C. Morris eds., *Andean Ecology and Civilization.* Univ. Tokyo Press, Tokyo.

Jurion, F. and J. Henry. 1969. *Can Primitive Farming be Modernized?* Pub. de L'Institut Nat. Pour L'Etude Agronomique du Congo. 457 pp.

Kahn, R. P. and J. L. Libby. 1958. The effect of environmental factors and plant age on the infection of rice by the blast fungus Pyricularia oryzae. Phytopathology 48: 25-30.

Kaiser, W. J. and G. M. Horner. 1980. Root rot of irrigated lentils in Iran. Canadian J. Bot. 58: 2549-2566.

Kaplan, L. 1960. Historical and ethnobotanical aspects of domestication in Tagetes. Econ. Bot. 14: 200-202.

Karani, P. K. 1986. Observations on the productivity of matoke (bananas) (Musa spp.) under agroforestry at Entebbe, Uganda. Commonw. For. Rev. 65: 241-251.

Karunairajan, R. 1982. Green manuring in the tropics. pp. 319-326. In: Hill, Stuart ed., *Basic Techniques in Ecological Farming.* Birkhäuser Verlag, Basel, Switzerland.

Kasasian, L. 1971. *Weed Control in the Tropics.* Leonard Hill, London. 307 pp.

Kass, D. C. L. 1978. *Polyculture Cropping Systems: Review and Analysis.* Cornell Intl. Agr. Bull. 32. Cornell Univ., Ithaca, NY. 69 pp.

Katan, J. 1981. Solar heating (solarization) of soil for control of soilborne pests. Annu. Rev. Phytopathol. 19: 211-236.

______. 1987. Soil solarization. pp. 77-105. In: Chet, I. ed., *Innovative Approaches to Plant Disease Control.* Wiley, New York.

Kayll, A. J. 1974. Use of fire in land management. pp. 483-511. In: Kozlowski, T. T. and C. E. Ahlgren. *Fire and Ecosystems.* Academic Press, New York.

Keitt, G. W. 1959. History of Plant Pathology. pp. 61-97. In: Horsfall, J. G. and A. E. Dimond eds., *Plant Pathology. An Advanced Treatise. Vol. 1. The Diseased Plant.* Academic Press, New York.

Kelman, A. and R. J. Cook. 1977. Plant pathology in the People's Republic of China. Ann. Rev. Phytopathol. 15: 409-429.

Kenneth, R. 1977. Sclerospora graminicola. pp. 96-99. In: Kranz, J., H. Schmutterer and W. Koch eds., *Diseases, Pests, and Weeds in Tropical Crops.* Verlag Paul Parey, Berlin.

Kerr, A. and W. R. F. Rodrigo. 1967. Epidemiology of tea blister blight (Exobasidium vexans). III. Spore deposition and disease prediction. Trans. Brit. Mycol. Soc. 50: 49-55.

Keswani, C. L. and Mreta, R. A. D. 1982. Effect of intercropping on severity of powdery mildew on green gram. pp. 110-114. In: Keswani, C. L. and B. J. Ndunguru, ed., Proc. 2nd Symp. on Intercropping in Semi-Arid Areas. IDRC - 186e, IDRC, Ottawa.

Kimber, C. T . 1973. Spatial patterning in the dooryard gardens of Puerto Rico. Geograph. Rev. 63: 6-26.

Kincaid, R. R. 1946. Soil factors affecting incidence of root knot. Soil Sci. 61: 101-109.

King, F. H. 1926. *Farmers of Forty Centuries.* Harcourt, Brace, New York. 379 pp.

King, L. D. 1990. Sustainable Soil Fertility Practices. pp. 144-177. In: Francis, C. A., C. Butler Flora and L. D. King. 1990. *Sustainable Agriculture in Temperate Zones.* Wiley, New York.

King, S. R. 1987. Four endemic Andean tuber crops: promising food resources for agricultural diversification. Mountain Res. and Develop. 7: 43-52.

Kirkby, R., P. Gallegos, and T. Cornick. 1980. Metodologia para el desarrollo de technologia agricola apropiada para pequeños productores, experiencias del Proyecto Quimiag-Penipe, Ecuador. Ministerio de Agric. y Ganaderia, Ecuador. 38 pp.

Knight, C. G. 1978. *Ecology and Change. Rural Modernization in an African Community.* Academic Press, New York. 300 pp.

Knowles, P. F. and M. D. Miller. 1965. Safflower. Calif. Agric. Exp. Sta. Ext. Serv. Circ. 532. Univ. of California, Davis, CA, 51 pp.

Kozaka, T. 1965. Control of rice blast by cultivation practices in Japan. pp. 421-438. In: IRRI. *The Rice Blast Disease.* Johns Hopkins Press, Baltimore.

Kozlowski, T. T. ed. 1984. *Flooding and Plant Growth.* Academic Press, New York. 356 pp.

______. ed. 1978. *Water Deficits and Plant Growth. Water and Plant Disease.* Vol. 5. Academic Press, New York. 323 pp.

Kranz, J., H. Schmutterer and W. Koch. eds. 1977. *Diseases, Pests, and Weeds in Tropical Crops.* Verlag Paul Parey, Berlin. 666 pp.

Kuan, T. L. and D. C. Erwin. 1980. Predisposition effect of water saturation of soil on Phytophthora root rot of Alfalfa. Phytopathology 70: 981-986

Kunstadter, P. E., E. C. Chapman, and S. Sabharsi eds. 1978. *Farmers in the Forest.* Univ. Hawaii Press, Honolulu. 402 pp.

Kupers, L. J. P. 1972. Problems in moncultures. EPPO Bull. 6: 107-120.

La Mondia, J. A. and B. B. Brodie. 1986. The effect of potato trap crops and fallow on decline of Globodera rostochiensis. Ann. Appl. Biol. 108: 347-352.

Lagemann, J. and J. Heuveldop. 1983. Characterization and evaluation of agroforestry systems: the case of Acosta-Puriscal, Costa Rica. Agroforestry

Systems 1: 101-115.
Lahman, L. K., M. D. Harrison, and M. Workman. 1981. Pre-harvest burning for control of tuber infection by Alternaria solani. Am. Potato J. 58: 593-599.
Lal, R. 1975. Role of mulching techniques in tropical soil and water management. IITA Tech. Bull. No. 1. 38 pp.
______. 1981. No-tillage farming in the tropics. pp. 103-151. In: Phillips, R.E., G. W. Thomas, and R. L. Blevins, ed., *No-tillage Research: Research Reports and Reviews.* Univ. Kentucky, Lexington, KE.
______. 1987. *Tropical Ecology and Physical Edaphology.* Wiley, New York. 732 pp.
Lampert, R. J. 1967. Horticulture in the New Guinea highlands -- C14 dating. Antiquity 41: 307.
Langford, M. H. 1945. South American Leaf Blight of Hevea Rubber Trees. U. S. Dept. of Agric. Tech. Bull. 882. 31 pp.
La Placa, P. J. and M. A. Powell. The agricultural cycle and the calendar at Pre-Sargonic Givsu. Bul. Sumerian Agriculture 5: 75-104.
Large, E. C. 1962. *The Advance of the Fungi.* Dover, NY. 488 pp.
Larios, J. F. and R. A. Moreno. 1977. Epidemiolgia de algunas enfermedades foliares de la yuca en diferentes sistemas de cultivo. II. Roya y muerte descendente. Turrialba 27: 151-156.
Laudelot, H. 1961. Dynamics of tropical soils in relation to their fallowing techniques. Paper 11266/E. FAO, Rome.
Leach, L. D. and R. H. Garber. 1970. Control of Rhizoctonia. pp. 189-198. In: Parmeter Jr., J. R. ed., *Rhizoctonia solani, Biology and Pathology.* Univ. Calif. Press, Berkeley, CA.
Leighty, C. E. 1938. Crop rotation. pp. 406-430. In: *Yearbook. U.S.D.A.,* Washington, DC.
Leihner, D. 1983. *Management and Evaluation of Intercropping Systems with Cassava.* CIAT, Cali, Colombia. 70 pp.
Lennon, T. J. 1982. Raised Fields of Lake Titicaca, Peru: A Pre-hispanic Water Management. Ph.D. Thesis. Univ. Colorado, Boulder, CO. 340 pp.
Leppik, E. E. 1970. Gene centers of plants as sources of disease resistance. Ann. Rev. Phytopathol. 8: 323-344.
Lewandowski, S. 1989. Three sisters -- a Iroquoian cultural complex. pp. 41-45. In: Barrerio, J. ed., *Indian Corn of the Americas. Gift to the World.* Northeast Indian Quarterly, American Indian Program, Cornell Univ., Ithaca, NY.
Lewis, C. D. 1941. *The Georgics of Virgil: A New Translation. Book 1.* Jonathan Cape, London.
Lewis, J. A. and G. C. Papavizas. 1975. Survival and multiplication of soil-borne plant pathogens as affected by plant tissue amendments. pp. 84-89. In: Bruehl, G. W. ed., *Biology and Control of Soil-Borne Pathogens.* Am. Phytopathol. Soc., St. Paul, MN.
Liebman, M. 1987. Polyculture Cropping Systems. pp. 115-125. In: Altieri, M. A. ed., *Agroecology. The Scientific Basis of Alternative Agriculture.* Div. of Biol. Control, Univ. Calif., Berkeley, CA.
Linderman, R. G. 1970. Plant residue decomposition products and their effect on

host roots and fungi pathogenic to roots. Phytopathology 60: 19-20.

Litsinger, J. A. and K. Moody. 1976. Integrated pest management in multiple cropping systems. pp. 293-316. In: Stelley, M. ed., *Multiple Cropping.* ASA Special Publ. 27, Madison, WI.

Longman, K. A. and J. Jenik. 1987. *Tropical Forest and its Environment.* Longman, London. 347 pp.

Loos, C. A. 1961. Eradication of the burrowing nematode, Radopholus similis, from bananas. Plant Dis. Reptr. 45: 457-461.

Lozano T., J. C. and E. R. Terry. 1976. Cassava diseases and their control. pp. 156-160. In: IDRC. Proc. 4th Symp. Soc. Trop. Root Crops. IDRC, Ottawa, Canada.

Lozano T., J. C., H. D. Thurston, G. E. Galvez E. 1969. Control del "Moko" del platano y banano causado por la bacteria Pseudomonas solanacearum. Agric. Trop. (Colombia) 25: 315-324.

Luc, M., R. A. Sikora, and J. Bridge. eds. 1990. *Plant Parasitic Nematodes in Subtropical and Tropical Agriculture.* CAB International, Kew, Surrey, UK., 629 pp.

Lumsden, R. D., J. A. Lewis, R. Garcia-E. and G. A. Frias. 1981. Suppression of pathogens in soils from traditional Mexican agricultural systems. Phytopathology 71: 891-892. (Abstr.)

Lumsden, R. D., R. García-E., J. A. Lewis, and G. A. Frias-T. 1987. Suppression of damping-off caused by Pythium spp. in soil from the indigenous Mexican chinampa agricultural system. Soil Biol. Biochem. 19: 501-508.

Luo, S. M and C. R. Han. 1990. Ecological agriculture in China. pp. 299-322. In: Edwards, C. A. et al, ed., *Sustainable Agricultural Systems.* Soil, Water Conservation Soc., Ankeny, Iowa.

Maas, P. W. T. 1967. Sigatoka disease -- incidence under Surinam climate conditions. pp. 74-76. In: Proc. Caribbean Food Crops Society. 5th Ann. Meeting, Vol. 5. Paramaribo, Surinam.

Madden, L. V., M. A. Ellis, G. G. Grove, K. M. Reynolds, and L. L. Wilson. 1991. Epidemiology and control of leather rot of strawberries. Plant Disease 75:439-446.

Maekawa, K. 1990. Cultivation methods in the Ur II period. Bul. Sumerian Agriculture 5: 115-145.

Mamami, M. 1978. El chuño: preparación, uso, almacenamiento. pp. 227-239. In: Ravines, R. 1978. *Tecnologiá Andina.* Instituto Estudios Peruanos, Lima.

Mangelsdorf, P. C. 1966. Genetic potentials for increasing yields of food crops and animals. pp. 66-71. In: *Prospects of the World Food Supply.* Symp. Proc. Nat. Acad. Sci., Washington, D.C.

Markim, F. L. 1943. Blueberry diseases in Maine. Maine Agric. Exp. Sta. Bull. 419. Univ. Maine, Orono, ME.

Marten, G. G. and P. Vityakon. 1986. Soil management in traditional agriculture. pp. 199-225. In: Marten, G. G. ed., *Traditional Agriculture in Southeast Asia.* Westview Press, Boulder, CO.

Martin, F. W., and R. M. Ruberte. 1975. *Edible Leaves of the Tropics.* Mayaguez Inst. Trop. Agric., Mayaguez, Puerto Rico. 235 pp.

Martin, H. and E. S. Salmon. 1931. The fungicidal properties of certain spray-fluids, VIII. The fungicidal properties of mineral, tar, and vegetable oils. J. Agr. Sci. 21: 638-658.

Mason, A. F. 1928. *Spraying, Dusting, and Fumigating of Plants.* Macmillan, New York. 539 pp.

Massey, R. E. 1931. Studies on blackarm disease of cotton. II. Emp. Cotton Growing Rev. 8: 187-213.

Masuda, S. ed. 1985. *Andean Ecology and Civilization: an Interdisciplinary Perspective on Andean Ecological Complementarity.* Univ. of Tokyo Press, Tokyo. 550 pp.

Mateos, P. F. ed. 1956. *Obras de Padre Barnabé Cobo de la Compania de Jesus.* Biblioteca de Autores Españoles, Atlas, Madrid. 952 pp.

Matheron, M. E. and S. M. Mircetich. 1985. Influence of flooding duration on development of Phytophthora root and crown rot of Juglans hindsii and paradox walnut rootstocks. Phytopathology 75: 973-976.

Mathewson, K. 1984. *Irrigation Horticulture in Highland Guatemala.* Westview Press, Boulder, CO. 180 pp.

Mayer, E. 1979. *Land-use in the Andes: Ecology and Agriculture in the Mantaro Valley of Peru with Special Reference to Potatoes.* Social Sci. Unit. CIP, Lima, Peru. 115 pp.

______. 1985. Production zones. pp. 45-84. In: Masuda, S. ed., *Andean Ecology and Civilization: an Interdisciplinary Perspective on Andean Ecological Complementarity.* Univ. Tokyo Press, Tokyo.

Mayer, E. and C. Fonseca M. 1979. *Sistemas Agrarios en la Cuenca del Rio Cañete.* Liberia Studium, Lima, Peru.

McCalla, T. M. and D. L. Plucknett. 1981. Collecting, transporting, and processing organic fertilizers. pp. 19-37. In: Plucknett, D. L. and H. L. Beemer, Jr, ed., *Vegetable Farming Systems in China.* Westview Press, Boulder, CO.

McLoughlin, Peter F. M. ed. 1970. *African Food Production Systems.* John Hopkins Press, Baltimore. 318 pp.

McWhorter, C. G. 1972. Flooding for Johnson-grass control. Weed Sci. 20: 238-241.

Meggers, B. J. 1971. *Amazonia. Man and Culture in a Counterfeit Paradise.* Aldine. Atherton, Chicago. 182 pp.

Meiklejohn, J. 1955. Effect of bush burning on the microflora of a Kenya upland soil. J. Soil Sci. 6: 111-118.

Merideth, D. S. 1970. Banana Leaf Spot (Sigatoka) Caused by Mycosphaerella musicola Lach. Commonw. Mycol. Inst., Kew, Surrey. Phytopathol. Paper No. 11. 147 pp.

Merrill, M. C. 1983. Eco-agriculture: a review of its history and philosophy. Biol. Agric. Hort. 1: 181-210.

Meyer de Schauensee, R. 1964. *The Birds of Colombia and Adjacent Areas of South and Central America.* Livingston, Narbeth, PA. 427 pp.

Michon, G., J. Bompard, P. Hecketswiller, and C. Ducatillon. 1983. Tropical forest architectural analysis as to agroforests in the humid tropics: the example of traditional village-agroforests in West Java. Agroforestry Systems 1: 117-129.

Michon, G., F. Mary, and J. Bompard. 1986. Multistoried agroforestry garden system in West Sumatra. Agroforestry Systems 4: 315-338.

Miller, D. E. and D. W. Burke. 1975. Effect of soil aeration on Fusarium root rot of beans. Phytopathology 65: 519-523.

Miller, P. R. 1953. Lowland rice culture as an economic control of Sclerotinia and root-knot nematodes. FAO Plant Protec. Bull. 1: 183-187.

Miller, R. B., J. D. Stout, K. E. Lee. 1955. Biological and chemical changes following scrub burning on New Zealand hill soil. N.Z.J. Sci. and Tech. Bull. 37: 290-313.

Milsum, J. N. and D. H. Grist. 1941. *Vegetable Gardening in Malaya.* Dept. Agric., Federated Malay States, Kuala Lumpur, Malaysia.

Mink, G. I. 1981. Control of plant diseases using disease-free stocks. pp. 317-346. In: Pimentel, D. ed., *Handbook of Pest Management in Agriculture.* CRC Press, Boca Raton, FL.

Miracle, M. P. 1967. *Agriculture in the Congo basin.* Univ. of Wisc. Press, Madison. 355 pp.

Mirocha, C. J. and C. M. Christensen. 1974. Fungus metabolites toxic to animals. Annu. Rev. Phytopathology 12: 303-330.

Mittleider, J. R. 1986. *Grow-bed gardening.* Woodbridge Press, Santa Barbara, CA. 200 pp.

Mizukami, T. and S. Wakimoto. 1969. Epidemiology and control of bacterial leaf blight of rice. Ann. Rev. Phytopathol. 7: 51-72.

Mohamed, R. and J. Teri. 1989. Mgeta farmers and bean diseases. ILEIA Newsletter 4 (3): 18-19.

Monares, A. 1979. Agro-economic evaluation of highland seed in the Cañete Valley, Peru. Social Sci. Unit Working Paper No. 1979-3. CIP, Lima, Peru. 14 pp.

Monastersky, R. 1988. Legacy of fire: the soil strikes back. Sci. News 133: 231.

Montoya M., J. M. and E. Schieber. 1970. La practica del doblado del maiz (Zea mays L.) y su relación con la incidencia de hongos en la mazorca. Turrialba 20: 24-29.

Monyo, J. H., A. D. R. Ker and M. Campbell. 1976. *Intercropping in Semi-Arid Areas.* IDRC 076e. IDRC, Ottawa, Canada. 72 pp.

Moody, K. 1975. Weeds and shifting cultivation. Pest Articles and News Sum. (PANS) 21: 188-194.

Moore, W. D. 1949. Flooding as a means of destroying the sclerotia of Sclerotinia sclerotiorum. Phytopathology 39: 920-927.

Mora, L. E. and R. A. Moreno. 1984. Cropping pattern and soil management influence on plant diseases: I. Diplodia macrospora leaf spot of maize. Turrialba 34: 35-40.

Moreno, R. A. 1975. Diseminacion de Ascochyta phaseolorum en variedades de frijol de costa bajo diferentes sistemas de cultivo. Turrialba 25: 361-364.

______. 1979. Crop protection implications of cassava intercropping. pp. 113-127. In: Weber, E., B. Nestel, and M. Campbell, ed., *Intercropping with Cassava.* Proc. Intl. Workshop, Trivandrim, India. 1978. IDRC 142-e. IDRC, Ottawa, Canada.

______. 1985. Plant pathology in the small farmer context. Ann. Rev.of

Phytopathol. 23: 491-512.
Moreno, R. A. and L. E. Mora. 1984. Cropping pattern and soil management influence on plant diseases: II. Bean rust epidemiology. Turrialba 34: 41-45.
Morley, S. G. and G. W. Brainerd. 1946. *La Civilización Maya.* Fondo Cultura Economica, Mexico. 527 pp.
Mountjoy, D. C. and S. R. Gliessman. 1988. Traditional management of a hillside agroecosystem in Tlaxcala, Mexico: an ecologically based maintenance system. Am. J. Alternative Agric. 3: 3-10.
Mt. Pleasant, J. 1989. The Iroquois Sustainers. Practices of a longterm agriculture in the Northeast. pp. 33-39. In: Barrerio, J. ed., *Indian Corn of the Americas. Gift to the World.* Northeast Indian Quarterly, American Indian Program, Cornell Univ., Ithaca, NY.
Mueller, S. C. and G. W. Fick. 1987. Response of susceptible and resistant alfalfa cultivars to Phytophthora root rot in the absence of measurable flooding damage. Agron. J. 79: 210-204
Muimba-Kankolongo, A., L. Simba, T. P. Singh and G. Muyolo. 1989. Outbreak of an unusual stem tip dieback of cassava (Manihot esculenta Crantz) in Western Zaire. Agric. Ecosystems and Environment 25: 151-164.
Mukiibi, J. K. 1982. Effects of intercropping on some diseases of beans and groundnuts. pp. 110-114. In: Keswani, C. L. and B. J. Ndunguru, ed., Proc. Second Symp. on Intercropping in Semi-Arid Areas. IDRC - 186e. IDRC, Ottawa
Muller, R. and P. S. Gooch. 1982. Organic amendments in nematode control. An examination of the literature. Nematropica 12: 319-326.
Muñoz, E. 1986. Producción de maíz, frijol, y calabaza en un sistema hidráulico de chinampa. (Maize, bean, and squash production in a chinampa irrigation system). Turrialba 36: 369-373.
Murdoch, W. W. 1975. Diversity, complexity, stability and pest control. J. Appl. Ecol. 12: 795-807.
Murphy, W. S., B. B. Brodie and J. M. Good. 1974. Population dynamics of plant nematodes in cultivated soil : effects of combinations of cropping systems and nematicides. J. Nematology 6: 103-107.
Murra, J. V. 1960. Rite and crop in the Inca State. pp. 393-407. In: Diamond, S. ed., *Culture in History.* Colombia Univ. Press, New York.
______. 1968. An Ayamara kingdom in 1567. Ethnohistory 15 (2): 115-149.
______. 1987. Una vision indigena del mundo Andino. pp. IL-LXIII. In: Poma de Ayala, Felipe Guaman. *Nueva Crónica y Buen Gobierno.* J. V. Murra, R. Adorno and J. L. Urioste eds., Historia 16, Madrid.
Museo Nacional de Culturas Populares. 1982. *Nuestro Maiz. Treinta Monografias Populares.* Secretaria Educación Publica, Mexico, D. F., Vol. 1. 327 pp., Vol. 2. 304 pp.
Mutsaers, H. J. W. 1978. Mixed cropping experiments with maize and grounduts. Netherlands J. Agric. Sci. 26: 344-353.
Myers, N. 1988a. Environmental degradation and some economic consequences in the Philippines. Environ. Conservation 15: 205-214.
______. 1988b. Threatened biotas: "hot spots" in tropical forests.

Environmentalist 8: 187-208.

______. 1988c. News and Comment: Tropical forests: shifting cultivators in the Philippines. Plants Today. January-February.

Nair, P. K. R. 1983. Agroforestry with coconuts and other tropical plantation crops. pp. 79-102. In: Huxley, P. A. ed., *Plant Research and Agroforestry*. ICRAF, Nairobi.

______. 1984. *Soil Productivity Aspects of Agroforestry*. Intl. Council Res. Agroforesty, Nairobi, Kenya. 85 pp.

Nair, P. K. R., R. Varma and E. V. Nelliat. 1974. Intercropping for enhanced profits from coconut plantation. Indian Farming 24 (4): 11-13.

Nataraj, T. and S. Subramanian. 1975. Effect of shade and exposure on the incidence of brown-eye-spot of coffee. Indian Coffee 39: 179-180.

National Acad. Sciences. 1968. *Plant-disease Development and Control*. Nat. Acad. Sciences, Washington, DC. 205 pp.

______. 1972. *Genetic Vulnerability of Major Crops*. Nat. Acad. Sciences, Washington, D.C. 307 pp.

National Research Council. 1978. *Postharvest Food Losses in Developing Countries*. Nat. Acad. Sciences. Washington, D.C. 206 pp.

______. 1982. Agriculture in the humid tropics. pp. 93-120. In: *Ecological Aspects of Development: Humid Tropics*. Nat. Acad. Press, Washington, DC.

______. 1989a. *Alternative Agriculture*. Nat. Acad. Sciences. Washington, D.C. 448 pp.

______. 1989b. *Lost Crops of the Incas: Little Known Plants of the Andes with Promise for Worldwide Cultivation*. Nat. Acad. Press, Washington, D.C. 415 pp.

Nations, J. D. 1988. The Lacandon Maya. pp. 86-88. In: Denslow, J. S. and C. Padoch eds., *People of the Tropical Rain Forest*. Univ. of Calif. Press, Berkeley, CA.

Navarro, L. 1903. La rabia (Ascochyta pisi Oud.) y la mosca de los garbanzales. Ministerio Agric., Industria, Commercio y Obras Publicas, Madrid. 95 pp.

Neergard, P. 1977. *Seed Pathology*. Vol. 1. Macmillan, London. 839 pp.

Nelliat, E. V. and N. K. Ji. 1976. Intensive cropping in coconut gardens. Indian Farming 26 (9): 9-12.

Nelson, M. R. 1989. Biological control: The second century. Plant Dis. 73: 616.

Nelson, R. R. 1973. *Breeding Plants for Disease Resistance. Concepts and Applications*. Pa. State Univ. Press, University Park, PA. 401 pp.

Nesmith, W. C. and S. F. Jenkins. 1985. Influence of antagonists and controlled matric potential on the survival of Pseudomonas solanacearum in four North Carolina soils. Phytopathology 75: 1182-1187.

Netting, R. M. 1968. *Hill Farmers of Nigeria. Cultural Ecology of the Kofyar of the Jos Plateau*. Univ. Washington Press, Seattle. 259 pp.

Newhall, A. G. 1955. Disinfestation of soil by heat, flooding, and fumigation. Bot. Rev. 21: 189-250.

Nickel, J. L. 1973. Pest situation in changing agricultural situations -- a review. Bul. Entomol. Soc. 19: 136-142.

Niñez, Vera K. 1984. Household gardens: Theoretical considerations on an old survival strategy. Report No. 1. Int. Potato Center (CIP), Lima, Peru. 41 pp.

______. 1985. Introduction: household gardens and small scale food production. Food and Nutrition Bull. 7 (3): 1-5.

Noble, M. and M. J. Richardson. 1968. *An Annotated List of Seed-borne Diseases.* Phytopathol. Papers No. 8. Commonwealth Mycol. Inst., Kew, Surrey, England. 191 pp.

Norgaard, R. B. 1984. Traditional agricultural knowledge: past performance, future prospects, and institutional implications. Am. J. Agric. Econ. 66: 874-878.

Norman, M. J. T. 1979. *Annual Cropping Systems in the Tropics: An Introduction.* Univ. Florida Press, Gainesville, FL. 276 pp.

Nusbaum, C. J. and H. Ferris. 1973. The role of cropping systems in nematode population management. Ann. Rev. Phytopathol. 11: 423-440.

Nye, P. H. and D. J. Greenland. 1960. *The Soil Under Shifting Cultivation.* Tech. Commun. 51. Commonwealth Bureau of Soils, Harpenden, England. 156 pp.

Nyoka, G. C. 1983. Potentials for no-tillage crop production in Sierra Leone. pp. 66-72. In: Lal, R. P. A. Sanchez, and R. W. Cummings, Jr, ed., *Land Clearing and Development in the Tropics.* Balkema, Rotterdam.

Ochse, J. J., M. J. Soule Jr., M. J. Dijkman, and C. Wehlburg. 1961. *Tropical and subtropical agriculture.* 2 Vols. Macmillan, New York. 1446 pp.

Ogbuji, R. O. 1979. Shifting cultivation discourages nematodes. World Crops 31 (3): 113-114.

Ogundana, S. K. , S. H. Z. Naqvi and J. A. Ekundayo. 1971. Studies on soft rot of yams in storage. Trans. Br. Mycological Soc. 56: 73-80.

Okafor, J. C. and E. C. M. Fernández. 1987. Compound farms in southeastern Nigeria: a predominant agroforestry home garden system with crops and small livestock. Agroforestry Systems 5: 153-168.

Okigbo, B. N. and D. J. Greenland. 1976. Intercropping systems in tropical Africa. pp. 63-101. In: Papendick, R. I., P. A. Sanchez and G. B. Triplett. eds., *Multiple Cropping.* Am.Soc. Agron. Special Publ. 27, Madison, WI.

Okigbo, B. N. and R. Lal. 1982. Residue mulches, intercropping and agri-silviculture potential in tropical Africa. pp. 54-69. In: Hill, S. ed., *Basic Techniques in Ecological Farming.* Birkhäuser Verlag, Basel, Switzerland.

Oldfield , M. L. 1984. *The Value of Conserving Genetic Resources.* U. S. Dept. Interior, Washington, DC. 360 pp.

Oldfield, M. L. and J. B. Alcorn. 1987. Conservation of traditional agroecosystems. BioScience 37: 199-208.

Oliveros, B., J. C. Lozano, and R. H. Booth. 1974. A Phytophthora root rot of cassava in Colombia. Plant Dis. Reptr. 58: 703-705.

Omohundro, J. T. 1985. One potato, two potato. Natural History 94 (6): 22-30.

Oomen, H. A. P. C., and G. J. H. Grubben. 1978. *Tropical Leaf Vegetables in Human Nutrition.* Royal Trop. Inst., Amsterdam. 140 pp.

Ordish, G. 1976. *The Constant Pest. A Short History of Pests and Their Control.* Scribner's, New York. 240 pp.

Orjuela, J. 1956. Factors affecting stripe rust of wheat in Colombia. Phytopathology 46: 22. (Abstr.)

Orlob, G. B. 1973. Ancient and Medieval Plant Pathology. Pflanzenschutz-

Nachrichten Bayer 26: 65-294.

Orlove, B. S. 1977. *Alpacas, Sheep and Men: the Wool Export Economy and Regional Society in Southern Peru.* Academic Press, New York. 270 pp.

Ou, S. H. 1972. *Rice Diseases.* Commonw. Mycol. Inst., Kew, England. 368 pp.

Padmanabhan, S. Y. 1973. The great Bengal famine. Ann. Rev. Phytopathol. 11: 11-26.

______. 1977. Cochliobolus miyabeanus. pp. 106-108. In: Kranz, J., H. Schmutterer and W. Koch, ed., *Diseases, Pests, and Weeds in Tropical Crops.* Verlag Paul Parey, Berlin.

Palte, J. G. L. 1989. Upland Farming on Java, Indonesia. Netherlands Geograph. Studies 97, Geograph. Inst., Univ. Utrecht, Netherlands. 250 pp.

Palti, J. 1981. *Cultural Practices and Infectious Crop Diseases.* Springer-Verlag, Berlin. 243 pp.

Paner, V. E. 1975. Multiple cropping research in the Philippines. pp. 182-202. In: Proc. Cropping Systems Workshop. IRRI, Los Baños, Philippines.

Papavizas, G. C. 1973. Status of applied biological control of soil-borne plant pathogens. Soil Biol. Biochem. 5: 709-720.

______. ed. 1981. *Biological Control in Crop Production.* Allanheld Osmun, London 461 pp.

Papavizas, G. C. and J. A. Lewis. 1979. Integrated control of Rhizoctonia solani. pp. 415-424. In: Schippers, B. and W. Gams, ed., *Soil-borne Plant Pathogens.* Academic Press, New York.

Papendick, R. I., P. A. Sanchez and G. B. Triplett. eds. 1976. *Multiple Cropping: Proceedings of a Symposium.* Am. Soc. Agron. , Madison, WI. 378 pp.

Parker, A. C. 1910. Iroquois uses of maize and other food plants. New York State Museum Bul. 144: 51-119.

Parker, J. M. 1989. Stability of Disease Expression in the Potato Late Blight Pathosystem. Ph.D. Thesis. Cornell Univ., Ithaca, NY. 141 pp.

Parsons, J. and W. M. Denevan. 1967. Pre-Colombian ridged fields. Sci. Am. 217 (1): 93-100.

Parsons, J. and N. P. Psuty. 1975. Sunken fields and preshispanic subsistence on the Peruvian coast. Am. Antiquity 40: 259-282.

Patiño, V. M. 1956. El maiz chococito. Noticia sobre su cultivo en America ecuatorial. America Indigena 16: 309-346.

______. 1962. El maiz chococito. Notas sobre su cultivo en America ecuatorial. Rev. Interamericana de Ciencias Sociales. 1 (3): 358-388.

______. 1965. *Historia de la Actividad Agropecuaria en America Equinocial.* Univ. del Valle. Imprenta Departmental, Cali, Colombia. 601 pp.

Patrick, Z. A., T. A. Toussoun, and L. W. Koch. 1964. Effect of crop-residue decomposition products on plant roots. Ann. Rev. Phytopathol. 2: 267-292.

Peters, W. J. and L. F. Neuenschwander. 1988. *Slash and Burn. Farming in the Third World Forest.* Univ. of Idaho Press, Moscow, ID. 113 pp.

Phillips, R. E. and S. H. Phillips. eds. 1984. *No-tillage Agriculture. Principles and Practices.* Van Nostrand Reinhold, New York, NY. 320 pp.

Phillips, R. E., R. L. Blevins, G. W. Thomas, W. W. Frye, and S. H. Phillips. 1980. No-tillage agriculture. Science 208: 1108-1113.

Pieczarka, K. J. and J. W. Lorbeer. 1974. Control of bottom rot of lettuce by ridging and fungicide application. Plant Dis. Reptr. 58: 837-840.

Pimentel, D. ed. 1981. *Handbook of Pest Management in Agriculture.* CRC Press, Boca Raton, FL. 597 pp.

Pinchinat, A. M., J. Soria, and R. Bazan. 1976. Multiple cropping in tropical America. pp. 51-61. In: Papendick, R. I., P. A. Sanchez and G. B. Triplett. eds., *Multiple Cropping.* ASA Special Publ. No. 27. Am. Soc. Agron., Madison, WI.

Plaisted, R. L., H. A. Mendoza, H. D. Thurston, E. E. Ewing, B. B. Brodie, and W. M. Tingey. 1987. Broadening the range of adaptation of andigena (neotuberosum) germplasm. CIP Circular 15 (2): 1-5.

Plucknett, D. L., N. J. H. Smith, J. T. Williams, and N. M. Anishetty. 1983. Crop germplasm conservation and developing countries. Science 220: 163-169.

Poltthanee, A. and G. G. Marten. 1986. Rainfed cropping systems in Northeast Thailand. pp. 103-131. In: Marten, G. G. ed., *Traditional Agriculture in Southeast Asia. A Human Ecology Perspective.* Westview, Boulder, Co.

Poma de Ayala, F. G. 1613. *Letter to a King: A Peruvian Chief's Account of Life Under the Incas and Under Spanish Rule.* Translated from *Nueva Cronica y Buen Gobierno* by C. Dilke. 1978. E. P. Dutton, New York. 248 pp.

______. 1987. *Nueva Cronica y Buen Gobierno.* J. V. Murra, R. Adorno and J. L. Urioste eds., Historia 16, Madrid. 1384 pp.

Ponnamperuma, F. N. 1972. The chemistry of submerged soils. Adv. Agron. 24: 29-96.

______. 1984. Effects of flooding on soils. pp. 9-45. In: Kozlowski, T. T. 1984. *Flooding and Plant Growth.* Academic Press, New York.

Porras, V. H., C. A. Cruz and J. J. Galindo. 1990. Manejo integrado de la mazorca negra y la moniliasis del cacao el el tropico húmedo bajo de Costa Rica. Turrialba 40: 238-245.

Posey, D. A. 1985. Indigenous management of tropical forest ecosystems: the case of the Kayapó Indians of the Brazilian Amazon. Agroforestry Systems 3: 139-159.

Prinz, D. and F. Rauch. 1987. The Bamenda model. Development of a sustainable land-use system in the highlands of Western Cameroon. Agroforestry Systems 5: 463-474.

Proctor, F. J., J. P. Goodliffe and D. G. Coursey. 1981. Post-harvest losses of vegetables and their control in the tropics. pp. 139-172. In: Spedding, C. R. W. ed., *Vegetable Productivity.* Macmillan, London.

Puleston, D. E. 1978. Terracing, raised fields, and tree cropping in the Maya lowlands: a new perspective in the geography of power. pp. 225-245. In: Harrison, P. D. and B. L. Turner II eds., *Pre-Hispanic Maya Agriculture.* Univ. New Mexico Press, Albuquerque.

Pullman, G. S., J. E. DeVay, R. H. Garber, and A. R. Weinhold. 1981. Soil solarization: effects on Verticillium wilt of cotton and soilborne populations of Verticillium dahliae, Pythium, spp., Rhizoctonia solani and Thielaviopsis basicola. Phytopathology 71: 954-959.

Purchase, I. F. H. ed. 1974. *Mycotoxins.* Elsevier, Amsterdam. 443 pp.

Purseglove, J. W. 1968. *Tropical crops: Dicotyledons.* Wiley, NY. 719 pp.

______. 1972. *Tropical Crops, Monocotyledons.* Wiley, NY. 607 pp.

Putter, C. A. J. 1980. The management of epidemic levels of endemic diseases under tropical subsistence farming conditions. pp. 93-103. In: Palti, J. and J. Kranz, ed., *Comparative Epidemiology.* PUDOC, Wageningen.

Rabinowitz, D., C. R. Linder, R. Ortega, D. Begazo, H. Murguia, D. S. Douches and C. F. Quiros. 1990. High levels of interspecific hybridization between Solanum sparsipilum and S. stenotomum in experimental plots in the Andes. Am. Potato J. 67: 73-81.

Rajaram, S. and A. Campos. 1974. Epidemiology of wheat rusts in the Western Hemisphere. CIMMYT Res. Bull. 27. CIMMYT, Mexico. 27 pp.

Raven, P. H. 1981. Tropical rain forests: A global responsibility. Natural History 90 (2): 29-32.

Ravines, R. 1978. *Tecnologiá Andina.* Instituto Estudios Peruanos, Lima. 821 pp.

______. 1980. *Chanchan. Metropoli Chimu.* Instituto Estudios Peruanos, Lima. 390 pp.

Raychaudhuri, S. P. ed. 1964. *Agriculture in Ancient India.* Indian Council of Agric. Res., New Delhi. 167 pp.

Recharte, J. 1989. Food and Money: Economic Culture Among the Peasant Gold Miners of the Cuyocuyo District (Puno, Peru). Ph.D. Thesis. Cornell Univ., Ithaca, NY.

Redclift, M. 1987. "Raised bed" agriculture in pre-Colombian Central and South America: a traditional solution to the problem of sustainable farming systems. Biol. Agric. Hort. 5: 51-59.

Reichel-Dolmatoff, G. 1965. *Colombia: Ancient People and Places.* Praeger, New York. 231 pp.

Rengifo-Vasquez, G. 1987. *La Agricultura Tradicional en los Andes. Manejo de Suelos, Sistemas de Labranza y Herramientas Agricolas.* Editorial Horizonte, Lima, Peru. 81 pp.

Repetto, R. C. 1985. *Paying the Price: Pesticide Subsidies in Developing Countries.* World Resources Inst., Washington, D. C. 27 pp.

Rhoades, H. L. 1964. Effect of fallowing and flooding on root-knot in peat soil. Plant Dis. Reptr. 48: 303-306

Rhodes, R. E. 1988. Thinking like a mountain. ILEIA Newsletter 4(1):3-5.

Rhoades, R., M. Benavides, J. Recharte, E. Schmidt, and R. Booth. 1988. *Traditional potato storage in Peru: Farmers' knowledge and practices.* Intl. Potato Center, Lima, Peru. Potatoes in Food Systems Res. Ser. Report No. 4. 67 pp.

Rice, D. S. 1991. The Maya redicovered. Roots. Nature 2/91: 10-14.

Rice, E. L. 1984. *Allelopathy.* Academic Press, New York. 422 pp.

Rich, A. E. 1983. *Potato Diseases.* Academic Press, New York. 238 pp.

Riley, T. and G. Freimuth. 1979. Field systems and frost drainage in the prehistoric agriculture of the Upper Great Lakes. Am. Antiquity 44 (2): 271-285.

Rocheleau, D., K. Wachira, L. Malaret, and B. M. Wanjohi. 1990. Local knowledge for agroforestry and native plants. pp. 14-24. In: Chambers, R. A. Pacey and L. A. Thrupp. 1990. *Farmer First. Farmer Innovation and*

Agricultural Research. Intermediate Technol. Publ., London. 219 pp.

Rockwood, W. G. and R. Lal. 1974. Mulch-tillage: a technique for soil and water conservation in the tropics. SPAN 17: 77-79.

Rodriguez-Kabana, R. 1986. Organic and inorganic nitrogen amendments to soil as nematode suppressants. J. Nematology 18: 129-135.

Rodriguez-Kabana, R., S. Jordan, J. Walfredo, and J. P. Hollis. 1965. Nematodes: biological control in rice fields: role of hydrogen sulfide. Science 148: 524-526.

Rodriguez-Kabana, R., and G. Morgan-Jones. 1987. Biological control of nematodes: soil amendments and microbial antagonists. Plant and Soil 100: 237-247.

Rodriguez-Kabana, R., and G. Morgan-Jones. 1988. Potential for nematode control by mycofloras endemic in the tropics. J. Nematology 20: 191-203.

Rohrbach, K. G. and W. J. Apt. 1986. Nematode and disease problems of pineapple. Plant Dis. 70: 81-87.

Rosado-May, F. J., R. Garcia-Espinosa and S. R. Gliessman. 1985. Impacto de los fitopatogenos del suelo al cultivo del frijol en suelos bajo differente manejo en la Chontalpa, Tabasco. Rev. Mex. Fitopatologia 3 (2): 80-90.

Rosado-May, F. J., and R. Garcia-Espinosa. 1986. Estrategias empiricas para el control de la mustia hilachoza (Thanatephorus cucumeris Frank Donk) de frijol comun en la Chontalpa, Tabasco. Rev. Mex. Fitopatologia 4: 109-113.

Rosado-May, F. J., S. R. Gliessman and M. Alejos Pedraza. 1986. Potencial alelopatico del cadillo (Bidens pilosa L.) y su relacion con el ataque de algunas fitopatogenos del suelo al maiz. Rev. Mex. Fitopatologia 4: 124-132.

Rosado-May, F. J. and S. R. Gliessman. 1988. Weed management as an alternative for controlling plant parasitic nematodes. pp. 515-523 In: Allen, P. and D. van Dusen, ed., *Global Perspectives on Agroecology and Sustainable Agricultural Systems*. Vol. 2. Agroecology Program, Univ. of California, Santa Cruz.

Ross, W. D. 1913. *The Work of Aristotle*. De Plantis. Oxford Univ. Press, Oxford.

Rotem, J. 1982. Modification of the plant canopy and its impact on plant disease. pp. 327-342. In: Hatfield, J. L. and I. J. Thomason. *Biometeorology in Integrated Pest Management*. Academic Press, New York.

Rowe, J. H. 1963. Inca culture at the time of the Spanish conquest. pp. 183-330. In: Steward, J. H. ed., *Handbook of South American Indians. Vol. 2. The Andean Civilizations*. Bur. Am. Ethnology Bull. 143. Cooper Square Publ., New York.

Rowe-Dutton, P. 1957. *The mulching of vegetables*. Commonw. Agric. Bur., Farnham Royal, England. 169 pp.

Ruddle, K. and G. Zhong. 1988. *Integrated Agriculture-Aquaculture in South China. The Dike-Pond System of the Zhujiang Delta*. University Press, Cambridge. 164 pp.

Ruiz Castro, A. 1944. *Enfermedades de la Vid*. Min. de Agricultura, Madrid. 252 pp.

Rupert, J. A. 1951. Rust resistance in the Mexican wheat improvement program. Foleto Tecnico No. 7. Oficina de Estudios Especiales, Mexico, D.F. Mexico. 44 pp.

Russo, R. O., and G. Budowski. 1986. Effect of pollarding frequency on biomass of Erythrina poeppigiana as a coffee shade tree. Agroforestry Systems 4: 145-162.

Ruthenberg, H. 1980. *Farming Systems in the Tropics.* Clarendon, Oxford. 424 pp.

Ruttan, V. W. 1988. Sustainability is not enough. Am. J. Alternative Agric. 3: 128-130.

Saari, E. E. and R. D. Wilcoxson. 1974. Plant disease situation of high-yielding dwarf wheats in Asia and Africa. Ann. Rev. Phytopathol. 12: 49-68.

Sagnia, S. 1989. Pest control in millet farming. ILEIA Newsletter 4 (3): 15-17.

Salas, H. 1988. Ecological reclamation of the Chinampa area of Xochimilco, Mexico City. pp. 469-473. In: Allen, P. and D. van Dusen eds., *Global Perspectives on Agroecology and Sustainable Agricultural Systems.* Vol. 2. Agroecology Prog., Univ. California, Santa Cruz, CA.

Salik, J. 1989. Ecological basis of Amuesha agriculture. Adv. Econ. Bot. 7: 189-212.

Salik, J. and M. Lundberg. 1990. Variation and change in Amuesha agriculture, Peruvian upper Amazon. Adv. Econ. Bot. 8: 199-223.

Sanchez, P. A. 1976. *Properties and Management of Soils in the Tropics.* Wiley, New York. 618 pp.

Sanchez, P. A. and J. R. Benites. 1987. Low-input cropping for acid soils of the humid tropics. Science 238: 1521-1527.

Sasser, J. N. 1971. An introduction to the plant nematode problem affecting world crops and a survey of current control methods. Pflanzenschutz-nachrichten 24: 3-47.

Sastrawinata, S. E. 1976. Nutrient Uptake, Insect, Disease, Labor Use, and Productivity Characteristics of Selected Traditional Intercropping Patterns Which Together Affect Their Continued Use by Farmers. Ph.D Thesis, College Agric., Univ. Philippines, Los Baños, Philippines.

Sattaur, O. 1988. The lost art of the waru waru. New Scientist 118 (1612): 50-51.

Sauer, C. O. 1952. *Agricultural Origins and Dispersals.* Am. Geograph. Soc., New York. 110 pp.

Sayre, R. M. 1971. Biotic influence in soil environment. pp. 235-256. In: Zuckerman, B. M., W. F. Mai, and R. A. Rhode eds., *Plant-Parasitic Nematodes.* Vol. 1. Academic Press, NY.

Schieber, E. 1977. Ceratocystis fimbriata. pp. 100-102. In: Kranz, J., H. Schmutterer, and W. Koch, ed., *Diseases, Pests and Weeds in Tropical Crops.* Verlag Paul Parey, Berlin.

_____. 1977. Mycena citricolor. pp. 160-161. In: Kranz, J., H. Schmutterer, and W. Koch, ed., *Diseases, Pests and Weeds in Tropical Crops.* Verlag Paul Parey, Berlin.

Schjellerup, I. 1985. Observations on ridged fields and terracing systems in the Northern Highlands of Peru. Tools and Tillage 5 (2): 100-121.

Schmidt, R. A. 1978. Diseases in forest ecosystems: The importance of functional diversity. pp. 287-315. In: Horsfall, J. G. and E. B. Cowling, ed., *Plant Disease - An Advanced Treatise. Vol. II. How Disease Develops in Populations.*

Academic Press, New York.

Schmutterer, H., K. R. S. Ascher and H. Rembold. eds. 1987. *Natural Pesticides from the Neem Tree (Azadirachta indica A. Juss).* GTZ, Eschborn, W. Germany. 297 pp.

Schnathorst, W. C. 1981. Life cycle and epidemiology of Verticillium. pp. 81-111. In: Mace, M. E., A. A. Bell, and C. H. Beckman. *Fungal Wilt Disease of Plants.* Academic Press, New York.

Schneider, R. W. ed. 1982. *Suppressive Soils and Plant Disease.* Am. Phytopthol. Soc., St. Paul, MN. 88 pp.

Schoeneweiss, D. F. 1975. Predisposition, stress, and plant disease. Ann. Rev. Phytopathol. 13: 193-211.

______. 1986. Water stress predisposition to disease -- an overview. pp. 157-174. In: Ayres, P. G. and L. Boddy, ed., *Water, Fungi and Plants.* Cambridge Univ. Press, Cambridge.

Schultz, J. L. 1974. Primitive and peasant economies. pp. 61-67 . In: Biggs, H. H. and R. L. Tinnermeir eds., *Small Farm Agricultural Development Problems.* Colorado State Univ., Fort Colliins, CO.

Schultz, T. W. 1964. *Transforming Traditional Agriculture.* Yale Univ. Press, New Haven. 212 pp.

Schwartz, H. F. and G. E. Galvez eds., 1980. *Bean Production Problems: Disease, Insect, Soil, and Climatic Constraints of Phaseolus vulgaris.* CIAT, Cali, Colombia. 434 pp.

Schwartz, H. F. and M. A. Pastor-Corrales. 1989. *Bean Production Problems in the Tropics.* CIAT, Cali, Colombia. 654 pp.

Seneviratne, S. N. de S. 1976. Bacterial wilt in solanaceous crops grown in rice fields. pp. 95-101. In: Sequeira, L. and A. Kelman, ed., Proc. First Intl. Planning Conference and Workshop on the Ecology and Control of Bacterial Wilt Caused by Pseudomonas solanacearum. July 18-23, 1976. North Carolina State Univ., Raleigh, N.C.

Serpenti, L. M. 1977. *Cultivators in the Swamps.* Van Gorcum, Assen, Amsterdam. 308 pp.

Sezgin, F. 1971. *Geschichte des Arabischen Schrifttums.* Vol. 4. Brill, Leiden. 398 pp.

Sharvelle, E. G. 1979. *Plant Disease Control.* Avi, Westport, CT. 331 pp.

Shea, S. R. and P. Broadbent. 1983. Developments in cultural and biological control of Phytophthora diseases. pp. 335-350. In: Erwin, D. C., S. Bartnicki-Garcia and P. Tsao eds., *Phytophthora. Its Biology, Taxonomy, Ecology and Pathology.* Am. Phytopathol. Soc., St. Paul, MN.

Shekhawat, G. S. , A. V. Gadewar, V. K. Bahal, and R. K. Verma. 1988. Cultural practices for managing bacterial wilt of potatoes. pp. 65-84. In: International Potato Center (CIP). Bacterial Diseases of the Potato. Report of Planning Conf. on Bacterial Diseases of the Potato. CIP, Lima, Peru.

Shepard, J. F. and L. E. Claflin. 1975. Critical analyses of the principles of seed potato certification. Ann. Rev. Phytopathol. 13: 271-293.

Shipton, P. J. 1977. Monoculture and soilborne plant pathogens. Ann. Rev. Phytopathol. 15: 387-407.

Siemens, A. H. 1980. Indicios de aprovechamiento agricola prehispanico de

tierras inundables en el centro de Veracruz. Biotica 5 (3): 83-92.

Siemens, A. H., and D. E. Puleston. 1972. Ridged fields and associated features in southern Campeche: new perspectives on the lowland Maya. Am. Antiquity 37: 228-239.

Simmonds, N. W. 1966. *Bananas*. Longman, London. 512 pp.

Simon, P. 1923. *Noticias Historiales de las Conquistas de Tierra Firme en las Indias Occidentales, Tomo 1, 2a Noticia, Cap. III.* Biblioteca Banco Popular, Bogota. 574 pp.

Skutch, A. 1950. Problems in milpa agriculture. Turrialba 1: 4-6.

Smith, A. E. and D. M. Secoy. 1975. Forerunners of pesticides in classical Greece and Rome. J. Agric. Food Chem. 23: 1050-1055.

Smith, A. E. and D. M. Secoy. 1985. Pest control methods practiced by British farmers during the 17th and 18th century. pp. 23-34. In: Fusonie, A. E. and M. L. Silva eds., *Searching Agriculture's Past.* J. Nat. Agric. Library Assoc. 10 (1/4). Jan. Dec.

Smith, E. C. S. 1985. A review of the relationship between shade types and cocoa pest and disease problems in Papua New Guinea. Papua New Guinea J. Agric., For., Fisheries 33: 79-88.

Smith, R. F. and H. J. Reynolds. 1972. Effects of manipulation of cotton agroecosystems on insect pest populations. pp. 373-406. In: Farvar, M. T. and J. P. Milton. eds., The Carless Technology. Natural History Press, New York. 1030 pp.

Soil Conservation Society of America. 1976. *Resource Conservation Glossary.* Soil Conservation Soc. of America, Ankeny, Iowa. 63 pp.

Soldi, Ana Maria. 1982. *La Agricultura Tradicional en Hoyas.* Fondo Editorial, Pontifica Univ. Catolica del Peru, Lima. 104 pp.

Soria, J., R. Bazan, A. M. Pinchinat, G. Paez, N. Mateo, R. Moreno, J. Fargas, and W. Forsythe. 1975. Investigación sobre sistemas de produción agricola para el pequeño agricultor del trópico. Turrialba 25: 283-283.

Spencer, J. E. 1966. *Shifting Cultivation in Southeast Asia.* Univ. California Press, Berkeley, CA. 247 pp.

Spencer, J. E. and G. A. Hale. 1961. The origin, nature, and distribution of agricultural terracing. Pacific Viewpoint 2 (1) : 1-40.

Sprauge, M. A. and G. L. Triplett eds. 1986. *No-tillage and Surface-tillage Agriculture. The Tillage Revolution.* Wiley, New York. 467 pp.

Spurr, M. S. 1986. *Arable Cultivation in Roman Italy c. 200 B.C.- c. A.D. 100.* Soc. Promotion Roman Studies, London. 159 pp.

Squier, E. G. 1858. *The States of Central America.* Harper, New York. 782 pp.

Sridhar, T. S. and S. Subramanian. 1969. Studies on the brown-eye spot of coffee. Indian Coffee 33 (3): 97-99.

Stadelman, R. 1940. *Maize culture in North Eastern Guatemala.* Carnegie Institution Publ. 523. Washington, D.C.

Stakman, E. C. and J. G. Harrar. 1957. *Principles of Plant Pathology.* Ronald Press, New York. 581 pp.

Steadman, J. R., D. P. Coyne, and G. E. Cook. 1973. Reduction of severity of white mold disease on Great Northern beans by wider row spacing and determinate growth habit. Plant Dis. Reptr. 57: 1070-1071.

Steiner, K. G. 1984. *Intercropping in Tropical Smallholder Agriculture with Special Reference to West Africa.* GTZ, Eschborn, W. Germany. 304 pp.

Stevens, F. L. 1932. Tropical plant pathology and mycology. Bull. Torey Bot. Club 59: 1-6.

Stevens, N. E. and I. Nienow. 1947. Plant disease control by unusual methods. Bot. Rev. 13: 116-124.

Stevens, R. B. 1974. *Plant Disease.* Ronald, New York. 459 pp.

Stevens, R. B. 1960. Cultural practices in disease control. pp. 357-429. In: Horsfall, J. G. and A. E. Diamond eds., *Plant Pathology. An Advanced Treatise.* Vol. 3. Academic Press, New York.

Stocks, A. 1983. Candoshi and Cocamilla swiddens in Easter Peru. Human Ecol. 11: 69-84.

Stoll, G. 1987. *Natural Crop Protection - Based on Local Farm Resources in the Tropics and Subtropics.* TRIOPS, Trop. Sci. Books, Langen, W. Germany. 188 pp.

Stolzy, L. H. and R. E. Sojka. 1984. Effects of flooding on plant disease. pp. 221-264. In: Kozlowski, T. T. ed., *Flooding and Plant Growth.* Academic Press, New York.

Stolzy, L. H., J. Letey, L. J. Klotz and C. K. Labanauskas. 1965. Water and aeration as factors in root decay of Citrus sinensis. Phytopathology 55: 270-275.

Stoner, W. N. and W. D. Moore. 1953. Lowland rice farming, a possible cultural control for Sclerotinia sclerotiorum in the Everglades. Plant Dis. Reptr. 37: 181-186.

Storey, H. H. 1936. Virus diseases of East African plants: VI. A progress report on the disease of cassava. E. Afr. Agric. J. 2: 34-39

Stover, R. H. 1954. Flood fallowing for eradication of Fusarium oxysporum f. cubense. II. Some factors involved in fungus survival. Soil Sci. 77: 401-414.

______. 1955. Flood fallowing for eradication of Fusarium oxysporum f. cubense III. Effect of oxygen on fungus survival. Soil Sci. 80: 397-412.

______. 1959. Growth and survival of root-disease fungi in soil. pp. 339-355 In: Holton, C.S. et al, ed., *Plant Pathology. Problems and Progress. 1908-1958.* Univ. Wisc. Press, Madison, WI.

______. 1962. *Fusarial Wilt (Panama Disease) of Bananas and other Musa Species.* Paper No. 4. Commonw. Mycol. Inst., Kew, Surrey. 316 pp.

______. 1979. Flooding of soil for disease control. pp. 19-28 In: Mulder, D. ed., 1979. *Soil Disinfestation.* Elseveir Sci. Publ., Amsterdam.

______. 1987. Plagas y Enfermedades de Caracter Epidemico en Cultivos Frutales de la Region Centroamericana. pp. 27-36. In: Pinochet, J. ed., Informe Technico No. 110. CATIE, Projecto MIP, Panama

Stover, R. H. and H. E. Ostmark. 1981. Integrated plant protection in bananas and plantains. pp. 564-567. In: Kommedahl, T. ed., *Proc. Symp. IX Intl. Congress of Plant Protection.* Vol II. Entom. Soc. Am., College Park, MD.

Straughan, B. 1991. The secrets of ancient Tiwanaku are benefiting today's Bolivia. Smithsonian 21 (11): 38-49.

Strømgaard, P. 1985. The infield-outfield system of shifting cultivation among the Bemba of South Central Africa. Tools and Tillage 5: 67-84.

______. 1988. The grassland mound-system of the Aisa-Mambwe of Zambia. Tools and Tillage 6: 34-46.

Sturhan, D. 1977a. Ditylenchus angustus. pp. 244-245. In: Kranz, J., H. Schmutterer, and W. Koch, ed., *Diseases, Pests and Weeds in Tropical Crops.* Verlag Paul Parey, Berlin.

______. 1977b. Radopholus similis. pp. 247-249. In: Kranz, J., H. Schmutterer, and W. Koch, ed., *Diseases, Pests and Weeds in Tropical Crops.* Verlag Paul Parey, Berlin.

Sturtevant, W. C. 1974. Woodsmen and Villagers of the East. pp. 101-149. In: Billard, J. B. *The World of the American Indian.* Natl. Geograph. Soc., Washington, D.C.

Su, Kuang-chi. 1979. Role of tomato in multiple cropping. In: Asian Vegetable and Develop. Center. 1979. Proc. 1st Intl. Symp. on Trop. Tomato. Oct. 23-27, 1978. Shanhua, Taiwan, Republic of China.

Suatmadjii, R. W. 1969. *Studies on the Effect of Tagetes Species on Plant Parasitic Nematodes.* H. Veenman & Zonen N. V., Wageningen, Netherlands. 132 pp.

Sumner, D. R., B. Doupnik Jr. and M. G. Boosalis. 1981. Effects of reduced tillage and multiple cropping on plant diseases. Ann. Rev. Phytopathol. 19:167-187.

Suneson, C. A. 1960. Genetic diversity -- a protection against plant diseases and insects. Agron. J. 52: 319-321.

Swanton, J. R. 1946. *The Indians of Southeastern United States.* Bur. Am. Ethnology, Bull. 137. Washington, D.C.

Tait, J. and B. Napopeth. eds. 1987. *Management of Pests and Pesticides. Farmer's Perceptions and Practices.* Westview, Boulder, CO. 244 pp.

Tapley, R. G. 1961. Crinkle-leaf of coffee in Tanganyika. 1961. Kenya Coffee 26: 156-157.

Tarr, S.A. J. 1972. *Principles of Plant Pathology.* Winchester Press, New York. 632 pp.

Teketay, D. 1990. Erythrina burana. Promising multipurpose tree from Ethiopia. Agroforestry Today 2: 13.

Teri, J. M. and R. A. Mohamed. 1988. Prospects for integrated disease management in the small farm context in Southern Africa. NorAgric Occasional Papers Series C. Development and Environment No. 3. Norwegian Center Intl. Agric. Development, Agric. Univ. Norway. Holstad, Norway. pp. 97-105

Terra, G. J. A. 1958. Farm systems in South-east Asia. Netherlands J. Agric. Sci. 6: 157-182.

Thames, Jr., W. H. and W. N. Stoner. 1953. A preliminary trial of lowland culture rice in rotation with vegetable crops as a means of reducing root-knot nematode infestations in Everglades. Plant Dis. Reptr. 37: 187-192.

Thomason, I. J. and P. Caswell. 1987. Principles of Nematode Control. pp. 87-130. In: Brown, R. H. and B. R. Kerry eds., *Principles and Practice of Nematode Control in Crops.* Academic Press, New York.

Thorne, G. 1961. *Principles of Nematology.* McGraw Hill, New York. 553 pp.

Thorold, C. A. 1940. Cultivation of bananas under shade for the control of leaf spot disease. Trop. Agric. (Trinidad) 17: 213-214.

Thresh, J. M. ed. 1981a. *Pests, Pathogens, and Vegetation. The Role of Weeds and Wild Plants in the Ecology of Crop Pests and Diseases.* Pitman, Boston. 517 pp.

______. 1981b. The role of weeds and wild plants in the epidemiology of plant virus diseases. pp. 53-70. In: Thresh, J. M. ed., *Pests, Pathogens, and Vegetation. The Role of Weeds and Wild Plants in the Ecology of Crop Pests and Diseases.* Pitman , Boston.

______. 1982. Cropping practices and virus spread. Ann. Rev. Phytopathol. 20: 193-218.

Thung, T. H. 1947. Potato diseases and hybridization. Phytopathology 37: 373-381.

Thurston, H. D. 1963. Bacterial wilt of potatoes in Colombia. Am. Potato J. 40: 381-390.

______. 1971. Relationship of general resistance: Late blight of potato. Phytopathology 61: 620-626.

______. 1980. International potato disease research for developing countries. Plant Dis. 64: 252-257.

______. 1984. *Tropical Plant Diseases.* Am. Phytopathol. Soc., St. Paul, MN. 208 pp.

______. 1989. *Enfermedades de Cultivos en el Tropico.* J. J. Galindo L. (trans.) Centro Agronómico Tropical de Investigación y Enseñanza (CATIE), Turrialba, Costa Rica. 232 pp.

______. 1990. Plant disease management practices of traditional farmers. Plant Disease 74: 96-102.

Thurston, H. D. and O. Schultz. 1981. Late blight. pp. 40-42. In: Hooker, W. J. ed., *Compendium of Potato Diseases.* Am. Phytopathol. Soc., St. Paul, MN.

Timothy, D. H. 1972. Plant germ plasm resources and utilization. pp. 631-666. In: Farvar, M. T. and J. P. Milton eds., *The Careless Technology. Ecology and International Development.* Natural History Press, Garden City, NY.

Todaro, M. P. 1977. *Economic Development in the Third World.* Longman, London. 588 pp.

Toussoun, T. A. and L. W. Koch. 1964. Effect of crop-residue decomposition products on plant roots. Ann. Rev. Phytopathol. 2: 267-292.

Treacy, J. M. 1989. Agricultural terraces in Peru's Colca Valley: promises and problems of an ancient technology. pp. 209-229. In: Browder, J. O. ed., *Fragile Lands of Latin America.* Westview, Boulder, CO.

Trenbath, B. R. 1974. Biomass productivity in mixtures. Adv. Agron. 26: 177-210.

______. 1976. Plant interactions in mixed crop communities. pp. 129-169. In: Papendick, R. I., P. A. Sanchez and G. B. Triplett, ed., *Multiple Cropping.* ASA Special Publ. No. 27. Am. Soc. Agron., Madison, WI.

______. 1986. Resource use by intercrops. pp. 57-81. In: Francis, C. A. ed., *Multiple Cropping Systems.* Macmillan, New York.

Trutmann, P., J. Voss, and J. Fairhead. in press. Disease control in bean cultivation systems of the Great Lakes region of Africa. Unpublished Manuscript..

Turner II, B. L. 1974. Prehistoric intensive agriculture in the Mayan lowlands.

Science 185: 118-124.

______. 1978a. The development and demise of the swidden thesis of Mayan agriculture. pp. 13-22. In: Harrison, P. D. and B. L. Turner II eds., *Pre-Hispanic Maya Agriculture.* Univ. New Mexico Press, Albuquerque.

______. 1978b. Ancient agricultural land use in the central Maya lowlands. pp. 163-183. In: Harrison, P. D. and B. L. Turner II eds., *Pre-Hispanic Maya Agriculture.* Univ. New Mexico Press, Albuquerque.

Turner II, B. L. and P. D. Harrison. 1981. Prehistoric raised-field agriculture in the Maya lowlands. Science 213: 399-405.

Ullstrup, A. J. 1972. The impacts of the southern corn leaf blight epidemics of 1970-1971. Ann. Rev. Phytopathol. 10: 37-50.

Upawansa, G. K. 1989. Ancient methods for modern dilemmas. ILEIA Newsletter 4 (3): 9-11.

Valencia, L. ed. 1986. *Control Integrado de Plagas de Papa.* CIP, Lima, Peru. 203 pp.

Valverde S., C. and D. E. Bandy. 1982. Production of annual food crops in the Amazon. pp. 243-280. In: Hecht, S. B. ed., *Amazonia. Agriculture and Land Use Research.* CIAT Ser. 03E (82). CIAT, Cali, Colombia.

Van Rheenen, H. A., O. E. Hasselbach and S. G. S. Muigai. 1981. The effect of growing beans together with maize on the incidence of bean diseases and pests. Neth. J. Plant Pathol. 87: 193-199.

Van Rheneen, M. A. 1979. Diversity of food beans in Kenya. Economic Botany 33: 448-454.

Vanderplank, J. E. 1963. *Plant Diseases: Epidemics and Control.* Academic Press, New York. 349 pp.

______. 1968. *Disease Resistance in Plants.* Academic Press, New York. 206 pp.

______. 1975. *Principles of Plant Infection.* Academic Press, New York. 216 pp.

Vandermeer, J. 1989. *The Ecology of Intercropping.* Cambridge Univ. Press, Cambridge. 237 pp.

Van Schreven, D. A. 1948. Investigations of certain pests and diseases of Vorstenlanden tobacco. Tijdschr. Plantenziekten 54: 149-174.

Varro, Marcus Terentius. 1934. *On Agriculture.* W. D. Hooper and H. B. Ash (trans.) Harvard Univ. Press, Cambridge. pp. 161-543.

Vasey, D. E. 1979. Population and agricultural intensity in the humid tropics. Human Ecol. 7: 269-283.

Vaughan, D. A. and L. A. Stich. 1991. Gene flow from the jungle to farmers. BioScience 41: 22-28.

Vavilov, N. I. 1951. *The origin, variation, immunity and breeding of cultivated plants.* Chronica Botanica 13: 1-366.

Vickers, W. T. 1983. Tropical forest mimicry in swiddens: a reassessment of Geertz's model with Amazonian data. Human Ecol. 11: 35-45.

Vidal y Cabasés, Francisco. 1778. *Conversaciones Instructivas en Que Se Trata de Fomentar la Agricultura por Medio del Riego de Las Tierras.* Imprenta Antonio de Sanchmaño, Madrid.

Villareal, R. 1980. *Tomatoes in the Tropics.* Westview Press, Boulder, CO. 174 pp.

Visser, T., N. Shanmuganathan, and J. V. Sabanayagam. 1961. The influence of sunshine and rain on tea blister blight, Exobasidium vexans Massee. Ann.

Appl. Biol. 49: 306-315.

von Hagen, V. W. ed. 1959. *The Incas of Pedro de Cieza de León.* Harriet de Onis (trans.) Univ. Oklahoma Press, Norman. 397 pp.

von Platen, H. H. 1985. *Appropriate Land Use Systems of Smallholder Farms on Steep Slopes in Costa Rica. A Study on Situation and Development Possibilities.* Wissenschaftsverlag Vauk, Kiel, Germany. 187 pp.

von Platen, H., P. Rodriguez, and J. Lagemann. 1982. *Farming systems in Acosta-Puriscal, Costa Rica.* CATIE, Turrialba, Costa Rica. 146 pp.

Waddell, E. 1972. *The Mound Builders: Agricultural Practices, Environment, and Society in the Central Highlands of New Guinea.* Univ. of Washington Press, Seattle, WA. 253 pp.

Wahl, I., Y. Anikster, J. Manisterski and A. Segal. 1984. Evolution at the center of origin. pp. 39-77. In: Bushnell, W. R. and A. P. Roelfs eds., *The Cereal Rusts.* Vol. 1. Academic Press, New York.

Walker, J. C. 1950. *Plant Pathology.* McGraw-Hill, New York. 699 pp.

Waller, J. M. 1977. Cercospora coffeicola. pp. 190-191. In: Kranz, J., H. Schmutterer, and W. Koch, ed., *Diseases, Pests and Weeds in Tropical Crops.* Verlag Paul Parey, Berlin.

Wang, J. 1983. *Taro*. Univ. Hawaii Press, Honolulu. 400 pp.

Warren, G. F. 1983. Technology transfer in no-tillage crop production in third world agriculture. pp. 25-31. In: Akobundu, I. O. and A. E. Deutsch, ed., *No-tillage Crop Production in the Tropics.* Proc. Symposium. Monrovia, Liberia. August 1981. Intl. Plant Protec. Center, Oregon State Univ., OR.

Wasilewski, A. 1987. The quiet epidemic. Pesticide poisoning in Asia. IDRC Reports 16: 16-17.

Watson, A. M. 1983. *Agricultural Innovation in the Early Islamic World. The Diffusion of Crops and Farming Techniques 700-1100.* Cambridge Univ. Press, Cambridge. 260 pp.

Weatherwax, P. 1954. *Indian Corn in Old America.* MacMillan, New York. 253 pp.

Weber, E., B. Nestel, and M. Campbell eds. 1979. *Intercropping with Cassava.* Proc. Intl. Workshop Held at Trivandrim, India, 27 Nov-1 Dec. 1978. IDRC-142e., IDRC, Ottawa, Canada. 144 pp.

Weber, G. F. 1939. Web-blight, a disease of beans caused by Corticum microsclerotia. Phytopathology 29: 559-575.

Webster, C. C. and P. N. Wilson. 1980. *Agriculture in the Tropics.* Longman, London. 640 pp.

Weeraratna, S. 1990. External inputs for sustainable agriculture. ILEIA Newsletter 6 (3): 20-21.

Wehunt, E. J. and Q. L. Holdeman. 1959. Nematode problems of the banana plants. Proc. Soil Crop Sci. Soc. Florida 19: 436-442.

Weinstock, J. A. 1984. Monoculture or polyculture in a swidden system. Human Ecol. 12: 481-482.

Wellman, F. L. 1938. Poverty of human requisites in relation to inhibition of plant diseases. Science 87: 64-65.

______. 1962. A few introductory features of tropical plant pathology.

Phytopathology 52: 928-930.

______. 1972. *Tropical American Plant Disease.* Scarecrow Press, Metuchen, NJ. 989 pp.

Werge, R. 1977. *Potato Storage Systems in the Mantaro Valley.* CIP, Lima, Peru.

______. 1979. Potato processing in the central highlands of Peru. Ecol. Food Nutrition 7: 229-234.

West, R. C. 1957. *The Pacific Lowland of Colombia: a Negroid Area of the American Tropics.* Louisiana State Univ. Studies. Social Sci. Ser. No. 8, La. State Univ. Press, Baton Rouge. 278 pp.

White, K. D. 1970. *Roman Farming.* Cornell Univ. Press, Ithaca, NY. 536 pp.

______. 1984. *Greek and Roman Technology.* Thames and Hudson, London. 272 pp.

Whitehead, A. G. 1977. Control of potato cyst-nematode, Globodera rostochiensis, Ro1, by picrolonic acid and potato trap crops. Ann. Appl. Biol 87: 225-227.

Wicks, T. and T. C. Lee. 1985. Effects of flooding, rootstocks and fungicides on Phytophthora crown rot of almonds. Australian J. Exp. Agric. 25: 705-710

Wiersum, K. F. 1983. Tree gardening and taungya on Java: examples of agroforestry techniques in the humid tropics. Agroforestry Systems 1: 53-70.

Wilcox, W.F. and S. M. Mircetich. 1985. Effects of flooding duration on the development of Phytophthora root and crown rots of cherry. Phytopathology 75: 1451-1455.

Wilken, G. C. 1969. Drained-field agriculture: an intensive farming system in Tlaxcala, Mexico. Geograph. Rev. 59: 215-241.

______. 1971. Food producing systems available to the ancient Maya. Am. Antiquity 36: 432:448.

Wilken, G. C. 1974. Some aspects of resource management by traditional farmers. pp. 47-59. In: Biggs, H. H. and R. L. Tinnermeir. *Small Farm Agricultural Development Problems.* Colorado State Univ., Fort Colliins, CO. 168 pp.

______. 1987. *Good Farmers. Traditional Agricultural Resource Management in Mexico and Central America.* Univ. of Calif. Press, Berkeley, CA. 302 pp.

Willey, R. W. 1975. The use of shade in coffee, cocoa, and tea. Hort. Abstracts 45: 791-798.

Williams, J. T. 1988. Identifying and protecting the origins of our food plants. pp. 240-247. In: E. O. Wilson. ed., *Biodiversity.* Nat. Acad. Press.

Williams, P. H. 1979. Vegetable crop production in the People's Republic of China. Ann. Rev. Phytopathol. 17: 311-324

______. 1981. Plant protection. pp. 129-162. In: Plucknett, D. L. and H. L. Beemer eds., *Vegetable Farming Systems in China.* Westview, Boulder, CO.

______. 1983. Conservation of plant and symbiont germplasm. pp. 131-144. In: Kommedahl, T. and P. H. Williams eds., *Challenging Problems in Plant Health.* Am. Phytopathol. Soc., St. Paul, MN.

Williams, R. J. 1984. Downy mildews of tropical cereals. pp. 1-103. In: Ingram, D. S. and P. H. Williams. ed., *Advances in Plant Pathology.* Vol. 2. Academic Press, New York.

Williams, R. J. and D. McDonald. 1983. Grain molds in the tropics: problems and

importance. Ann. Rev. Phytopathol. 21: 153-178.
Wilson, E. O. ed. 1988a. *Biodiversity*. Nat. Acad. Press, Washington, D. C. 521 pp.
______. 1988b. The current state of biological diversity. pp. 3-18. In: Wilson, E. O. ed., *Biodiversity*. Nat. Acad. Press, Washington, D. C.
Wilson, G. F. and K. L. Akapa. 1983. Providing mulches for no-tillage in the tropics. pp. 51-65. In: Akobundu, I. O. and A. E. Deutsch, ed., *No-tillage Crop Production in the Tropics.* Symp. Monrovia, Liberia, August 1981. IPPC, Oregon State Univ., Corvallis,OR.
Wilson, G. F. and F. E. Caveness. 1980. Effect of rotation crops on survival of root-knot, root-lesion, and spiral nematodes. Nematropica 10: 56-61.
Wilson, G. L. 1987. *Buffalo Bird Woman's Garden.* Agriculture of the Hidatsa Indians. Minnesota Historical Soc. Press, St. Paul. 129 pp.
Wilson, S. G. 1941. Agricultural practices among the Angoni-Tumbuka tribes of Mzimba (Nyasaland). E. Afr. Agric. J. 7: 89-93.
Wiseman, F. M. 1978. Agricultural and historical ecology of the Maya lowlands. pp. 63-115. In: Harrison, P. D. and B. L. Turner II eds., *Pre-Hispanic Maya Agriculture.* Univ. of New Mexico Press, Albuquerque, NM.
Witter, E. and J. M. Lopez-Real. 1987. The potential of sewage sludge and composting in a nitrogen recycling strategy for agriculture. Biol. Agric. Hort. 5: 1-23.
Wittwer, S., Y. Youtai, S. Han and W. Lianzheng. 1987. *Feeding a Billion: Frontiers of Chinese Agriculture.* Michigan State Univ. Press. 462 pp.
Wrigley, G. 1981. *Tropical Agriculture. The Development of Production.* Longman, London. 496 pp.
______. 1988. *Coffee*. Longman, London. 639 p.
Wrightson, J. 1889. *Fallow and Fodder Crops.* Chapman and Hall, London. 276 pp.
Yang, R. Z. and C. S. Tang. 1988. Plants used for pest control in China: a literature review. Econ. Bot. 42: 376-406 .
Yarwood , C. E. 1951. Defoliation by a rain-favored, a dew-favored, and a shade favored disease. Phytopathology 41: 194-195.
______. 1978. Water and the infection process. pp. 141-173. In: Kozlowski, T. T. ed., *Water Deficits and Plant Growth. Water and Plant Disease.* Vol. 5. Academic Press, New York.
Yen, D. E. 1974a. *The Sweet Potato and Oceania. An Essay in Ethnobotany.* Bishop Museum Press, Honolulu. 389 pp.
______. 1974b. Arboriculture in the subsistence of Santa Cruz, Solomon Islands. Econ. Bot. 28: 247-284.
Yoshii, K. and F. Varon de Agudelo. 1977. Los efectos de Tagetes minuta y Crotalaria spectabilis en la población de nematodos y en el rendemiento posterior de la soya. Fitopatologia 12: 15-19.
Young, H. M. 1982. *No-tillage Farming.* No-Till Farmer Inc., Brookfield, WI.
Youtai, Y. 1987. Agricultural history over seven thousand years. pp. 19-33. In: Wittwer, S., Y. Youtai, S. Han and W. Lianzheng, ed., *Feeding a Billion: Frontiers of Chinese Agriculture.* Michigan State Univ. Press.
Zadoks, J. C. and R. D. Schein. 1979. *Epidemiology and Plant Disease Management.*

Oxford, Univ. Press, Oxford. 427 pp.
Zehrer, W. 1986. Traditional agriculture and integrated pest management. ILEIA Newsletter No. 6. pp. 4-6.
Zinke, P. J., S. Sabhasri, and P. Kunstadter. 1978. Soil fertility aspects of the Lua forest fallow system of shifting cultivation. pp. 134-159 In: Kunstadter, P. E., E. C. Chapman, and S. Sabharsi eds., *Farmers in the Forest*. Univ. Hawaii Press, Honolulu.
Zuckerman, B. M., M. B. Dicklow, G. C. Coles, R. Garcia E., and N. Marban-Mendoza. 1989. Suppression of plant parasitic nematodes in the chinampa agricultural soils. J. Chemical Ecol. 15:1947-1955.

Disease Index

Pathogen Index

General Index

About the Book and Author

Most scientists and many of the world's farmers have abandoned traditional farming practices and systems in an effort to increase production and to improve the efficiency of land and labor use. The resulting "modern" systems largely ignore many of the sustainable pest management practices that have evolved among farmers over centuries. In this book, H. David Thurston catalogs and reviews valuable practices that are in danger of being lost in the modernization process.

Ancient farmers developed sustainable agricultural practices, including disease management, that allowed them to produce food and fiber for thousands of years with few outside inputs. Most systems were developed empirically through millennia of trial and error, natural selection, and observation. These proven practices often conserved energy, maintained natural resources, and reduced chemical use. A high level of diversity contributed to making traditional systems stable, resilient, and efficient. Thurston evaluates the sustainability, labor requirements, and external inputs needed for these diverse systems and their management, providing a comprehensive summary of effective traditional agricultural practices for plant disease management.

H. David Thurston is professor of international agriculture and plant pathology at Cornell University.